Epigenetics and Health

Epigenetics and Health

A Practical Guide

Michelle McCulley
Otago University
New Zealand

Published by John Wiley & Sons, Inc., Hoboken, New Jersey.
Published simultaneously in Canada.

For general information on our other products and services or for technical support, please contact our Customer Care Department within the United States at (800) 762-2974, outside the United States at (317) 572-3993 or fax (317) 572-4002.

Wiley also publishes its books in a variety of electronic formats. Some content that appears in print may not be available in electronic formats. For more information about Wiley products, visit our website at www.wiley.com.

Library of Congress Cataloging-in-Publication Data
Names: McCulley, Michelle, author.
Title: Epigenetics and health : a practical guide / Michelle McCulley.
Description: Hoboken, New Jersey : Wiley, [2024] | Includes bibliographical
   references and index.
Identifiers: LCCN 2023049009 (print) | LCCN 2023049010 (ebook) | ISBN
   9781119307983 (paperback) | ISBN 9781119307990 (adobe pdf) | ISBN
   9781119308003 (epub)
Subjects: MESH: Epigenomics
Classification: LCC RB155 (print) | LCC RB155 (ebook) | NLM QU 476 | DDC
   616/.042–dc23/eng/20231215
LC record available at https://lccn.loc.gov/2023049009
LC ebook record available at https://lccn.loc.gov/2023049010

Cover image(s): © SCIEPRO/SCIENCE PHOTO LIBRARY/Getty Images;
LAGUNA DESIGN/Getty Images; ArtemisDiana/Shutterstock
Cover design: Wiley

Set in 9.5/12.5pt STIXTwoText by Straive, Pondicherry, India

Contents

Preface

Epigenetics is the study of heritable changes in gene expression or cellular phenotype that do not involve changes to the underlying DNA sequence. These changes can be influenced by a variety of factors, including environmental factors, diet and lifestyle, and can have a significant impact on an individual's health and development. The importance of epigenetics lies in the fact that it provides a mechanism for the inheritance of traits that are not coded for in the DNA sequence itself. Epigenetic modifications can be passed down from one generation to the next and can influence the expression of genes that are involved in a wide range of biological processes, including development, ageing and disease.

Understanding the role of epigenetics is particularly important in the study of complex diseases such as cancer and neurological disorders, as it provides a potential avenue for developing new diagnostic and therapeutic strategies. By identifying specific epigenetic changes associated with these diseases, researchers can develop targeted treatments that can help to mitigate their effects and improve patient outcomes. Overall, the study of epigenetics is essential for gaining a deeper understanding of the complex interplay between genetics and environmental factors in shaping an individual's health and development and has the potential to revolutionise our understanding of human biology and disease.

This book will explore selected epigenetic phenomena as applied to human health and disease. We will investigate how, through epigenetic mechanisms, our genome is responsive to a wide range of environmental influences including nutrition, toxins and social circumstances. The mechanisms controlling these effects and their phenotypic outcomes will be covered. By the end of the book, the reader should understand the differences between genetic and epigenetic influences on gene expression, the range of epigenetic mechanisms used to regulate gene expression, how epigenetic modifications are propagated, and the phenotypic consequences for health and disease. In this book, the reader will rapidly build their knowledge to develop their understanding of epigenetics to a point where they can apply their learning to formulate a research question that relates to their own specific interest in this contemporary field.

Overview

The first three chapters of this book are designed to think about health issues from an epigenomic perspective, to understand molecular homeostasis and the interplay between genes and the environment. Chapter 4 focuses on tissues, methods, and resources for analysis. Chapters 5–7 focus on specific research areas; the concluding chapter focuses on future research questions and how the epigenome is a target for health and medicine. We end the book with a practical step-by-step guide to planning an epigenomics research project.

1

How Do Genes Work?

In this chapter, we will discuss fundamental concepts central to molecular biology and to understanding epigenetics. We will briefly discuss DNA, genes and proteins, mutation, genotype and phenotype and the relationships between each, and finally provide an overview as to how genes are controlled, outlining gene regulation and repression. We will outline the concept of molecular homeostasis and the interaction between genome and environment, briefly differentiating between epigenetics and genetics.

1.1   DNA, Genes and Proteins

DNA is comprised of four nucleotide bases: adenine, guanine, cytosine and thymine. DNA is a stable double helix with guanine always pairing with cytosine and adenine always pairing with thymine, or uracil in the case of RNA. The human genome is diploid, so it is comprised of 2 sets of 23 chromosomes, 1 inherited from each parent. Every nucleated cell contains the same 46 chromosomes with the same genetic information – this is our genome. Approximately 10% of the human genome is estimated to be coding, that is, specifically encodes for proteins; the remainder is noncoding DNA including repetitive sequences such as microsatellites, minisatellites, transposable elements (SINES, LINES), satellite DNA and triplet repeats.

Type of region	Description
Protein-coding genes	Segments of DNA that encode proteins.
Noncoding RNA genes	Segments of DNA that encode RNA molecules that do not translate into proteins. Examples include microRNAs and long noncoding RNAs.
Regulatory regions	DNA sequences that control gene expression, such as promoters, enhancers, silencers and insulators. These regions can be located upstream or downstream of the gene, or even within the gene itself.
Repetitive DNA	Segments of DNA that are repeated many times throughout the genome, such as transposable elements, satellite DNA and minisatellites.
Intergenic regions	DNA sequences located between genes that do not have any known function. These regions make up the majority of the human genome.
Introns	Segments of DNA located within genes that do not encode proteins. Introns are transcribed into RNA but are removed during the process of RNA splicing, which generates the final mRNA transcript that is used to produce proteins.

The human genome project suggests that there are around 20 000 genes, with each human chromosome on average containing 1300 genes. Genes are transcribed and translated to produce their encoded protein. This two-stage process takes the message embedded in double-stranded DNA and transcribes the coding sequence as single-stranded mRNA, which is then translated by the ribosome using tRNA and rRNA to produce the protein. Genes range in size from 1 kb, as in the case of insulin, to 2.5 Mb for larger genes, such as dystrophin. Almost all genes contain introns and exons. Exons are expressed coding regions and introns are non-expressed intervening sequences. Introns are transcribed into primary RNA and then spliced out of mature RNA in the cytoplasm. The average number of exons for a human gene is 9, and therefore introns are 8. However, there is considerable variation, e.g. 79 for dystrophin and 3 for beta-globin. The average size of an exon is 145 bp.

In addition to introns and exons, genes also have an adjacent upstream (5′) regulatory promoter region as well as other regulatory sequences such as enhancers, silencers and sometimes a locus control region. The promoter region contains specific conserved sequences such as TATA box, CG box and CAAT box, which provide binding sites for transcription factors. The first and last exons also contain untranslated regions (UTRs) known as the 5′ UTR and 3′ UTR. The 5′ UTR signals the start of transcription and contains ATG, the initiator codon that initiates the site of the start of translation. The 3′ UTR contains a termination codon, which marks the end of translation, plus nucleotides that encode a sequence of adenosine residues known as the poly (A) tail; the addition of a poly (A) tail is an essential step in the process of transcription that enables the pre-mRNA to exit the nucleus and move into the cytoplasm for translation.

Gene regulatory sequence	Function
Promoters	DNA sequences located upstream of the transcription start site that recruit the transcriptional machinery to initiate gene expression
Enhancers	DNA sequences that can be located far away from the gene they regulate and can enhance gene expression by increasing transcription rates and/or making expression more specific to certain cells or conditions
Silencers	DNA sequences that can be located near or far from the gene they regulate and can reduce or turn off gene expression
Insulators	DNA sequences that act as boundaries between different gene regulatory regions, preventing their influence on each other
Scaffold/matrix attachment regions	DNA sequences anchor the chromatin to the nuclear matrix, thereby organising and stabilising the structure of chromatin and regulating gene expression
CpG islands	Regions of DNA that have a high density of CpG dinucleotides are often associated with gene promoters and are involved in the regulation of gene expression through DNA methylation
miRNA target sites	Specific RNA sequences within the 3′ UTR of messenger RNAs that are recognised by microRNAs and lead to translational repression or mRNA degradation
Cis-regulatory modules	Clusters of enhancers, silencers and promoter sequences that work together to regulate the expression of a specific gene or set of genes in response to different signalling pathways

Pre-mRNA contains the entire transcribed sequences of exons and introns; the introns must be removed to produce mRNA that comprises the precise code for translation into a functional protein product. The removal of introns occurs through splicing, which occurs in a spliceosome, itself composed of hundreds of proteins and 5 RNAs. Once a transcript has been spliced and its 5′ and 3′ ends modified, a piece of mature functional mRNA has been produced, suitable for translation into a protein. There are specific nucleotide recognition sequences that aid in splicing. These sequences are present: towards the end of an exon, at the

Promoter   Transcribed region   Terminator

↓   ↓   ↓

5'--(regulatory elements)-[---intron---]----exon----[---intron---]----exon----(stop codon)-3'

Figure 1.1 The gene is shown in a linear arrangement, with the promoter at the beginning, the transcribed region in the middle and the terminator at the end. The transcribed region is shown with a series of exons (which code for protein) separated by introns (which do not code for protein). The regulatory elements, which can be located upstream or downstream of the promoter, are also shown at the beginning of the gene. At the end of the transcribed region, there is a stop codon that signals the end of the protein-coding sequence.

beginning of an intron, at the end of an intron, at the beginning of the next exon and at a region within the intron but close to the 3' end, which provides a binding site for intron removal. Such sequences are recognised by small nuclear ribonucleoproteins (snRNPs), and they cut the RNA at the intron-exon borders and connect the exons together. Alternative splicing is a mechanism that allows more information to be packed into a single gene. That is, from a single gene, through splicing exons together in different combinations, multiple RNAs and functional proteins can be produced; this allows cells to produce related but distinct proteins from a single gene. For example, one type of protein may be produced in one tissue, whereas another form may be produced in another tissue (Figure 1.1).

The estimated 20 000 protein-coding genes comprising the human genome are spread between the lengths of the human chromosomes. Each gene has a promoter where transcription of mRNA by RNA polymerase II is initiated by transcription factors. There are also remote elements called enhancers that will modulate the activity of the promoter. Different cell types activate gene expression in a different manner by making use of the genome's extensive system of regulatory *cis*-acting elements. The mechanisms underlying this process involve physical changes to the chromatin that either promote or inhibit gene expression. This is fundamental to understanding epigenetics, as it is how gene expression is suppressed or enhanced.

1.2 Gene Control, Homeostasis and Epigenetics

In complex multicellular organisms, differential gene expression is fundamental during embryonic development and in the maintenance of the adult state. It is key to understand that different cells make different proteins, which means different genes are switched on in different tissues even though all cells carry the same comprehensive set of genetic instructions. Therefore, there must be a way in which the body controls which gene is switched on, to what extent and when. Unused genetic information is not discorded – just not switched; for this to happen, there are specific mechanisms that can activate specific portions of the genome and repress the expression of other genes. The activation and repression of genetic loci can be seen as a form of molecular homeostasis, a delicate balancing act for a healthy organism, given expression of the wrong gene at the wrong time in the wrong cell type or in the wrong amount can lead to a harmful phenotype even when the gene itself is normal, such as in cancer or cell death.

Housekeeping genes need to be expressed in all types of nucleated cells because they encode a vital product needed to fulfil an important cellular function, for example, protein synthesis, energy production. Many other genes, however, show a much more restricted pattern of expression that is tissue specific. Spatial restriction of gene expression can occur at many distinct levels: multiple organ/tissue, specific tissue/cell lineage/cell type, individual cells and intracellular distribution. Similarly, gene expression is also temporally restricted at various levels: cell cycle stage, developmental stage, differentiation stage and inducible expression.

Gene expression requires two events to occur: first, the recruitment and activation of chromatin remodelling enzymes that alter the structure of the nucleosome and make the promoter sites on the DNA accessible. The second is the recruitment of coactivators to help assemble the factors needed for transcription; this includes transcription factors, RNA polymerase II, etc. Most eukaryotic genes are regulated at the transcriptional level either via the complex interplay between transcription factors, gene promoters and enhancers and/or through epigenetic mechanisms including chromatin remodelling and DNA methylation. Posttranscriptional gene regulation is through the regulation of splicing and mRNA processing, regulation of mRNA transport, degradation of mRNA, translational regulation or by modifying the translated protein to alter its activity.

Chromatin remodelling is essential for many processes including transcription, replication, DNA repair and recombination. Many of these changes are mediated by chemical changes to the N-terminal tails of core histones in a nucleosome. The histone tails are exposed on the surface of the nucleosome and are amenable to modification through acetylation, phosphorylation and methylation, allowing access to the underlying DNA. Acetylation adds an acetyl group to specific lysine residues; acetylated histones are found in regions of open chromatin where transcription occurs, which allows transcriptional enzymes to have access to the DNA. Acetylation is mediated by histone acetyltransferases (HATS) and deacetylation by histone deacetylases (HDACs) – both important to gene expression.

1.3 DNA Methylation and Regulation of Gene Expression

The DNA of most eukaryotes is modified after replication through the addition of methyl groups to bases and sugars. Base methylation most often involves enzyme-mediated addition of methyl groups to cytosine; in most eukaryotes, approximately 5% of cytosine residues are methylated; however, the extent of methylation can be tissue specific and can vary from less than 2% to over 7%. In eukaryotes, low amounts of methylation are associated with elevated levels of gene expression and elevated levels of methylation are associated with low levels of gene expression. In mammalian females, the inactivated X chromosome has a higher level of methylation than the active X chromosome. Methylation patterns are tissue specific and, once established, heritable for all cells of that tissue. The precise mechanism as to how methylation affects gene regulation is still uncertain. It is possible that there are proteins that bind to the methyl group attached to the cytosine; these proteins could recruit corepressors or HDACs to remodel chromatin, changing it from open to closed.

1.4 Post-transcriptional Regulation of Gene Expression

As outlined above, primary mRNA resulting from transcription is then further modified prior to translation – noncoding introns are removed, exons are spliced together and mRNA is modified through 5′ cap and poly A tail, prior to export to the cytoplasm for translation. Whilst many opportunities exist for further regulation during these steps, the major two are alternative splicing and regulation of the stability of the mRNA itself. Alternative splicing can generate multiple forms of a protein, so the expression of one gene can produce a family of related proteins.

Post-translational regulation	Mechanism	Examples
Protein phosphorylation	Addition of phosphate group to serine, threonine or tyrosine residues	Activation of Cyclin-dependent kinase 1 during cell cycle progression
Protein methylation	Addition of methyl group to arginine or lysine residues	Histone methylation by EZH2 in Polycomb repressive complex
Protein acetylation	Addition of acetyl group to lysine residues	Histone acetylation by histone acetyltransferases (HATs) in chromatin remodelling
Protein ubiquitination	Addition of ubiquitin to lysine residues	Ubiquitination of tumour suppressor p53 leading to its degradation
Protein sumoylation	Addition of small ubiquitin-like modifier (SUMO) protein to lysine residues	SUMOylation of transcription factor Sp3 leading to repression of its activity

Post-translational regulation	Mechanism	Examples
Protein neddylation	Addition of NEDD8 to lysine residues	Neddylation of Cullin proteins leading to activation of Cullin-RING E3 ubiquitin ligases
Protein glycosylation	Addition of carbohydrate to serine, threonine or asparagine residues	Glycosylation of E-cadherin promoting its adhesive function in cell–cell junctions
Protein lipidation	Addition of lipid group to cysteine residue	Palmitoylation of Hedgehog protein for proper signalling in development

1.5 Promoters and Enhancers

Transcription in eukaryotes is controlled through the interactions of promoters and enhancers; there are other localised sequence-based elements such as the CCAAT box, which also have an important regulatory effect and are found closer to the gene itself within the promoter region. Enhancers control the rate of transcription and can be located before, after or within the gene expressed. Transcription factors are proteins that bind to DNA-recognition sequences within promoters and enhancers and then activate transcription through protein–protein interaction. A gene can have different methylation patterns in different tissues and will be correlated with different regulatory patterns.

As mentioned previously, there are two types of regulatory sequences that control gene transcription: promoters and enhancers. Promoters are the recognition point for RNA polymerase binding; they are located immediately next to a gene, typically several hundred nucleotides long and need to be able to allow binding of RNA polymerase II to transcribe primary mRNA. The promoter region has several key elements that aid in its recognition by the polymerase enzyme; the promoter itself – the TATA box (made of 8-bp consensus sequence only AT base pairs), typically flanked either side by GC-rich regions. Many promoters also contain CAAT box and GC box- both bind transcription factors and function like enhancers- mutations in either can affect the rate of transcription. RNA polymerase II requires transcription factors to aid in the start of transcription; they are assembled at the promoter in a specific order and provide a platform that RNA polymerase II can recognise and bind to. The organisation of the upstream region of a gene including the promoter region is variable with respect to the nature, number and arrangement of controlling elements, in some cases including sites for enhancer binding and tissue-specific enhancer binding.

Enhancers can be on either side of a gene, so upstream (5′) or downstream (3′), they can also be at a reasonable distance from the gene or even within the gene itself. If they are adjacent to the gene, they are termed *cis*-regulators, as opposed to *trans* regulators such as binding proteins, which can regulate a gene on any chromosome. Typically, enhancers will interact with multiple regulatory proteins and transcription factors to increase the rate of transcription efficiency or promoter activation. Within enhancer sites, it is common to find binding sites for both positive and negative gene regulators. The main difference between enhancers and promoters is that enhancers do not have a fixed position (i.e., can be upstream, downstream or within the gene they regulate); the orientation of the enhancer can be inverted without significant effect on its action, and if an unrelated gene is placed near an enhancer, the transcription of that gene becomes enhanced. Enhancers are responsible for time- and tissue-specific expression. They exert their effect when the transcription factor binds to the enhancer, altering the chromatin configuration, and second, through binding/looping DNA, they bring distant enhancers and their promoters into direct contact to form complexes with transcription factors and polymerases. The new resulting configuration increases the overall rate of RNA synthesis.

1.6 Mutation Genotype, Phenotype, Epigenotype

Now we understand the structure and function of a gene, which is to carry instructions to produce a specific protein, we can begin to understand what the consequences are of mutations to DNA and disruptions to normal control over gene expression, that is, the switching on and off of a gene. In the simplest terms, the DNA spanning a particular area of a chromosome will contain the information needed to produce a particular protein in response to a signal to do so. If there are disruptions or changes to the DNA sequence of a particular gene, then this can have an impact on how or even whether the protein is produced. Given the complexity of transcription and translation, including the splicing of introns, exons and mechanisms that control gene expression, there are multiple locations where this can have a detrimental effect in terms of influencing the production of a protein. The resulting physical/physiological effect or the effect on a person's health is termed the phenotype; the change to the DNA is termed the genotype. There are also changes to the gene control mechanisms external to the DNA; these are termed epigenotypes.

What then is the significance of mutation? Are all mutations equal? Fundamental to molecular pathology is trying to understand why a particular genetic/epigenetic change or genotype should result in a specific phenotype or clinical condition. When trying to understand mutation, we can take a tiered approach, with the first stage being to ascertain whether the mutation results in loss of function or gain of function. With a loss of function, the protein product will have reduced or no functional capability. With a gain of function mutation, the protein product will do something atypical or unusual. Deletions of a whole gene, nonsense mutations and frameshifts are almost certain to destroy gene function. Mutations that change conserved sequence flanking introns affect splicing and will typically knock out the function of the gene; splicing can be changed by many other sequence changes too. Missense mutation is more likely to be pathogenic if it affects a part of the protein that is known to be functionally important, for example, a DNA-binding domain. Changing an amino acid is more likely to affect function if that changed amino acid happens to be conserved in related genes. Amino acid substitutions are more likely to affect function if they are non-conservative.

We will first look at the main classes of DNA mutations, which are deletions, insertions, frameshifts, dynamic mutations and single base substitutions (including missense mutations, nonsense mutations and splice site mutations). If we look at these from the perspective of how you stop a gene from working, then it can help to determine what meaningful interpretations of such data are. Genetic variants can also have an impact on epigenetic mark; single SNPs can have an impact on more than one proximal CpG sites. Correlation between SNPs and methylation is referred to as mQTLs – methylation quantitative trait loci. The added complexity is that SNPs can affect both gene expression and methylation independently, as outlined in Table 1.1.

1.7 Conclusion

Virtually any disease is the result of the combined action of genes and the environment, but the relative role of the genetic component may be large or small. For many of the common diseases affecting humans, such as diabetes, heart disease and cancer, there is not a single gene responsible. We are complex organisms, and as such, our pathophysiology is often a combined effect of what we inherit and our individual environment or circumstances.

There are many contributory factors to human variation both genetic, environmental and the interaction between the two. In terms of genetic differences, this is the existence of alternative forms of a gene (alleles) in the population. These variations can occur at greater or lesser frequency in a population, can be well established from a distant ancestor or can be new recently occurring variants. The impact of any change in genetic code depends on the resulting impact that change has on the production and activity of the encoded protein. Therefore, it is commonplace for similar phenotypes to arise as a consequence of mutation and variation at different loci.

Table 1.1 How do you stop a gene from working?

How to disrupt gene function	Example
Delete entire gene or part of the gene	
Insert a sequence into the gene	Normal gene sequence:
Disrupt the gene structure	5′--ATGCGTTAACCGGTACCG--3′
• By translocation	Met-Arg-Asn-Arg-Thr
• By inversion	
Prevent the promoter working	Mutated gene sequence:
• By mutation	5′--ATGCGTTCACCGGTACCG--3′
• By methylation	Met-Arg-His-Arg-Thr
Destabilise mRNA	
Polyadenylation site mutation	In this example, a single nucleotide mutation has changed the third codon from 'AAC' to 'CAC', which results in the substitution of the amino acid asparagine (Asn) with histidine (His). This change can potentially affect the function of the protein product of this gene, depending on its role and the location of the mutated residue within the protein structure. If the mutation occurs in a critical part of the protein, it may disrupt its activity or stability, leading to a loss of function. Alternatively, if the mutation occurs in a non-critical region, the protein may still be functional, but its activity may be altered or reduced. In some cases, mutations can also create new or altered functions, which can have both positive and negative consequences.
Nonsense-mediated RNA decay	
Prevent correct splicing	
• Inactivate donor splice site	
• Inactivate acceptor splice site	
• Alter an exonic splicing enhancer	
• Activate a cryptic splice site	
Introduce a frameshift in translation	
Convert a codon into a stop codon	
Replace an essential amino acid	
Prevent post-transcriptional processing	
Prevent correct cellular localisation of product	

This table summarises the many ways in which gene function can be disrupted.

Genes and their encoded protein products do not act in isolation; most genes will have multiple roles and functions, often depending on the cell type, tissue and developmental stage of the organism. There are many complex mechanisms that regulate when and how a gene is transcribed and translated into a protein and how the protein product is then modified and contributes to a particular biochemical pathway. Understanding human health and disease is therefore incredibly complex and far more than a simple genotype/phenotype correlation, with a need to further our understanding of the importance of gene–gene and gene–environmental interactions in disease.

In addition to mutation, epigenetic modifications and miRNAs can enhance or suppress gene expression. In the following chapter, we will explore further DNA methylation, the post-translational modifications of histones, chromatin remodelling and non-coding RNAs and their roles in controlling gene expression. As we have already discussed, these epigenetic mechanisms are intricately connected to developmental plasticity and cellular differentiation and being a means to relay environmental influences to the cell nucleus, bridging the gap between lifestyle and the genome.

Task

Questions to ask yourself about your research

- *Is epigenomics relevant to my research?*
- *How is your research interest an epigenetic/epigenomic issue? Or how is it not?*

Task 1. Make a mind map of known contributing factors to your disease/health phenotype of interest. Think about the condition, the associated phenotypes, the organism you are interested in studying, the time course of the disease and known contributory environmental and genetic factors. Think about whether there are differences in terms of gender, age, ethnicity and occupation. Think about how environmental and lifestyle factors contribute to the aetiology; what is known about the epigenetics of your condition?

Look at the obesity foresight map to see an incredibly detailed map.
Link to foresight map: https://assets.publishing.service.gov.uk/government/uploads/system/uploads/attachment_data/file/296290/obesity-map-full-hi-res.pdf

Further Reading

Tompkins JD. Discovering DNA Methylation, the history and future of the writing on DNA. *J Hist Biol*. 2022 Dec;55(4):865–87. doi: 10.1007/s10739-022-09691-8. Epub 2022 Oct 14. PMID: 36239862; PMCID: PMC9941238.

This paper explores the history and future of the discovery of DNA methylation. The author provides a comprehensive overview of the key scientists and research breakthroughs that have contributed to our understanding of DNA methylation, including the discovery of 5-methylcytosine in the 1950s and the development of techniques for mapping DNA methylation patterns in the genome. The article also discusses the potential applications of DNA methylation research in fields such as epigenetics, cancer research and personalised medicine. The author concludes that continued research in this field has the potential to transform our understanding of the role of DNA methylation in human health and disease and to lead to the development of new diagnostic and therapeutic tools.

Choudhuri S. From Waddington's epigenetic landscape to small noncoding RNA: some important milestones in the history of epigenetics research. *Toxicol Mech Methods*. 2011 May;21(4):252–74. doi: 10.3109/15376516.2011.559695. PMID: 21495865.

This paper provides an historical overview of the field of epigenetics, tracing its origins from Conrad Waddington's concept of the 'epigenetic landscape' in the 1940s to the discovery of small noncoding RNA in the 1990s. The author discusses key milestones in the development of the field, including the discovery of DNA methylation in the 1970s and the identification of histone modifications and chromatin remodelling complexes in the 1980s. The paper also highlights some of the key challenges facing the field of epigenetics, including the need for more comprehensive and systematic analyses of epigenetic marks and their functional consequences.

Van Soom A, Peelman L, Holt WV, Fazeli A. An introduction to epigenetics as the link between genotype and environment: a personal view. *Reprod Domest Anim*. 2014 Sep;49(Suppl 3):2–10. doi: 10.1111/rda.12341. PMID: 25220743.

In this article, the authors introduce the field of epigenetics and its role as a link between genotype and environment. The article first defines epigenetics and its various mechanisms, including DNA methylation, histone modification and non-coding RNAs. The authors then discuss how epigenetic modifications can be influenced by environmental factors, such as nutrition, stress and toxins, and how these modifications can have downstream effects on gene expression and disease susceptibility. The article concludes by highlighting the importance of understanding epigenetic regulation in the context of reproductive biology and animal breeding, as well as in the field of human health and disease.

Rivera RM, Bennett LB. Epigenetics in humans: an overview. *Curr Opin Endocrinol Diabetes Obes*. 2010 Dec;17(6):493–9. doi: 10.1097/MED.0b013e3283404f4b. PMID: 20962634.

This paper provides an overview of epigenetics in humans. They discuss the different types of epigenetic marks, such as DNA methylation, histone modifications and non-coding RNA regulation, and their impact on gene expression. The authors also describe how epigenetics is involved in various processes, including development, ageing and disease. They discuss the role of environmental factors, such as diet and stress, in shaping the epigenome and influencing health outcomes. Lastly, the authors touch on the potential clinical applications of epigenetics, such as epigenetic biomarkers for disease diagnosis and treatment.

Peixoto P, Cartron PF, Serandour AA, Hervouet E. From 1957 to nowadays: a brief history of epigenetics. *Int J Mol Sci.* 2020 Oct 14;21(20):7571. doi: 10.3390/ijms21207571. PMID: 33066397; PMCID: PMC7588895.

This paper provides a brief history of the field of epigenetics from 1957 to the present day. The authors discuss the key discoveries and milestones in the field, including the identification of histone modifications, DNA methylation and the role of non-coding RNAs in epigenetic regulation. The article also highlights the importance of epigenetics in various biological processes, including development, ageing and disease. The authors conclude by discussing the current state of epigenetics research and the potential for epigenetic therapies for disease treatment and prevention.

Bošković A, Rando OJ. Transgenerational epigenetic inheritance. *Annu Rev Genet.* 2018 Nov 23;52:21–41. doi: 10.1146/annurev-genet-120417-031404. Epub 2018 Aug 30. PMID: 30160987.

This article focuses on transgenerational epigenetic inheritance, which refers to the transfer of epigenetic information across multiple generations in the absence of changes to the underlying DNA sequence. The authors describe various mechanisms of transgenerational epigenetic inheritance, including DNA methylation, histone modification and small RNA-mediated processes, and highlight the role of these mechanisms in various model organisms. They also discuss the potential impact of environmental exposures on transgenerational epigenetic inheritance and the implications for human health and disease. The authors suggest that further research in this area is necessary to fully understand the mechanisms underlying transgenerational epigenetic inheritance and its potential implications.

2

What is Epigenetics?

Epigenetics refers to the study of heritable changes in gene expression that occur without any alteration in the underlying DNA sequence. These changes can be influenced by various environmental factors, such as diet, stress and exposure to toxins. Epigenetic modifications can occur at different levels, including DNA methylation, histone modification and non-coding RNA (ncRNA) expression, and can result in altered gene expression patterns that can persist across generations. Epigenetic modifications can play a critical role in development, ageing and disease susceptibility, and understanding the underlying mechanisms of epigenetic regulation is an active area of research in biology and medicine.

An important function of epigenetic processes is purely data management; that is, ensuring the genetic information is suitably packaged in the nucleus of a cell. The human genome comprises approx. 3 billion bps organised into 23 chromosomes. Each diploid cell has 46 chromosomes and therefore contains 6 billion bps of DNA, each base measures 0.34 nm, therefore in every nucleus of every diploid cell there must be 2 m of condensed DNA. The average person has approx. 70 trillion cells. Histones have an integral role in organising long strands of DNA as chromatin. Histone proteins H1, H2A, H2B, H3 and H4 are involved in compacting the DNA; 147 bps DNA is wrapped around dimers of histones H2A, H2B, H3 and H4 making up the nucleosome core. We will continue to develop ideas around gene regulation and the interaction between environment and genome in this chapter, where we will discuss more specifically epigenetic mechanisms, specifically methylation, histone modifications and other known epigenetic mechanisms (miRNA, polycomb, prions). Whilst different epigenetic mechanisms are to be discussed in this chapter, the focus for the remainder of the book will be centred on DNA methylation.

The epigenome comprises a range of modifications to DNA, RNA and proteins (histones). It is controlled structurally and functionally by specific enzymes responsible for catalysing the reactions to lay down these epigenomic marks and

Epigenetics and Health: A Practical Guide, First Edition. Michelle McCulley.
© 2024 John Wiley & Sons, Inc. Published 2024 by John Wiley & Sons, Inc.

Table 2.1 Types of epigenetic modifications.

DNA methylation changes: CpG site Methylation (hypermethylation and hypomethylation)	Epigenetic changes that involve the addition or removal of a methyl group to a cytosine base within a CpG dinucleotide.
Histone modification changes: Acetylation, Methylation, Phosphorylation, Ubiquitination	Epigenetic changes that involve modifications to the N-terminal tails of histone proteins, such as acetylation, methylation, phosphorylation and ubiquitination.
Chromatin remodelling changes: ATP-dependent Chromatin remodelling	Epigenetic changes that involve changes to the structure and packaging of chromatin, such as ATP-dependent chromatin remodelling.
Non-coding RNA changes: MiRNA, lncRNA	Epigenetic changes that involve alterations to non-coding RNAs, such as microRNAs (miRNAs) and long non-coding RNAs (lncRNAs). These can affect gene expression by regulating the stability or translation of messenger RNA (mRNA).

another set of enzymes responsible for removing both active and repressive marks to enable the epigenome to have plasticity and reversibility. The epigenomic marks are important in modulating gene expression; they are interpreted by other proteins and provide instructions, as to the activity of the gene or genes in the locality of the mark (Table 2.1).

Epigenetics focuses on heritable and potentially reversible changes to gene expression, which are not related to changes in the genomic DNA sequence. Mechanisms that regulate the transcription or gene expression levels that are cell-type or tissue-specific without altering the DNA. These mechanisms are biochemical modifications including the addition of a methyl group to cytosines as well as the post-translational modifications of histones. These mechanisms have a key role in cellular development, particularly in processes such as embryogenesis, cell differentiation, X inactivation and genomic imprinting.

2.1 Properties and Functions of Heterochromatin and Euchromatin

Chromatin can be described as either low-density euchromatin or high-density heterochromatin; post-translational modifications of the histone proteins are crucial in the maintenance of each structure. Heterochromatin and euchromatin represent two distinct chromatin states with different structural and functional properties. These two states play critical roles in regulating gene expression, maintaining chromosome integrity and organising the genome. Heterochromatin is compact, transcriptionally repressive and involved in maintaining genome stability. Euchromatin is less compact, associated with active gene expression and plays a crucial role in regulating gene transcription and other cellular processes. The interplay between these two chromatin states is essential for maintaining a functional and adaptable genome.

DNA packaging around histones has a direct relationship with gene expression; variable gene expression is linked to human health and disease phenotypes. Heterochromatin is condensed and makes it hard for transcription enzymes to access the DNA. Heterochromatin is tightly packed and densely stained under a microscope, indicating its condensed structure and contains highly repetitive DNA sequences, often referred to as 'junk DNA'. Heterochromatin is generally transcriptionally inactive or repressed due to its compacted structure and helps maintain the structural integrity of chromosomes. It prevents interactions between repeated sequences that could lead to chromosomal rearrangements and helps silence transposable elements, the mobile genetic elements that can cause genomic instability, preventing their movement and potential damage to the genome. Heterochromatin is also crucial for the proper function of centromeres and telomeres, playing a key role in centromere assembly and kinetochore formation, as well as protecting telomeres from degradation and fusion. In female humans, one X chromosome is inactivated to achieve gene dosage balance between males and females. The inactive X chromosome forms a condensed structure resembling heterochromatin, known as a Barr body.

The transition between heterochromatic and euchromatic states is determined through interactions between DNA and histone proteins. The histone proteins become modified post-translation via mechanisms such as methylation, which increases the interaction between the DNA and the histones, or acetylation, which tends to loosen interaction. During replication, the DNA and histones do not interact in this manner to facilitate the working of the polymerase enzymes. Euchromatin has a less compacted structure and is less densely stained under a microscope, indicative of its more open and accessible nature, making it permissive to gene transcription. It contains the majority of protein-coding genes and actively transcribed regions of the genome that are accessible to transcription factors and other regulatory molecules, allowing for mRNA synthesis and protein production. Euchromatin is therefore associated with gene expression and contains regulatory elements like promoters and enhancers. These elements control the timing and levels of gene expression by interacting with transcription factors and other regulatory proteins. Euchromatin regions are more amenable to changes in gene expression in response to developmental signals, environmental cues and cellular needs.

DNA is stored folded in the nucleus of the cell, extracting genetic information depending on the extent of coiling/uncoiling of the DNA to allow accessibility of transcriptional machinery. The relaxing and tightening of the chromatin is tightly linked to the regulation of gene expression. For gene expression to occur, transcriptional activators need to bind to specific upstream DNA sequences in gene promoter or enhancer sequences. Generally speaking, there are gene-specific activators that recruit chromatin remodelling factors, histone chaperones and histone acetyl-transferase complexes that loosen the core promoter chromatin around the TATA-box and the transcriptional start site to allow the formation of the preinitiation complex comprising general transcription factors FIIA, TFIIB, TFIID, TFIIE, TFIIF, TFIIH, and RNA polymerase II, adjacent to the +1 nucleosome.

In summary, heterochromatin and euchromatin represent two distinct chromatin states with different structural and functional properties. Heterochromatin is compact, transcriptionally repressive and involved in maintaining genome stability. Euchromatin is less compact, associated with active gene expression and plays a crucial role in regulating gene transcription and other cellular processes. The interplay between these two chromatin states is essential for maintaining a functional and adaptable genome.

2.2 DNA Methylation

DNA methylation is a type of epigenetic modification that involves the addition of a methyl group to the cytosine nucleotides in DNA, typically at CpG dinucleotides. This modification is catalysed by DNA methyltransferases (DNMTs) and can have a profound effect on gene expression by altering the accessibility of DNA to transcription factors and other regulatory proteins. In general, DNA methylation is associated with repression of gene expression, particularly when it occurs in promoter regions or near gene regulatory elements. Methylation of CpG islands in promoter regions can prevent the binding of transcription factors, leading to reduced transcriptional activity. In addition, methylated DNA can recruit proteins that promote the formation of repressive chromatin structures, further limiting gene expression. However, DNA methylation can also have a positive effect on gene expression in certain contexts. For example, methylation of enhancer regions can promote the binding of transcription factors and enhance gene expression. Moreover, DNA methylation can be dynamically regulated, allowing for rapid changes in gene expression in response to environmental cues.

The addition of a methyl group to the 5′ carbon of cytosine in a CpG dinucleotide. Generally associated with decreased transcription of the gene because of decreased binding of transcription factors to the gene promoter binding/enhancer site. The human genome comprises approximately 30 million CpG dinucleotides, which exist in one of three forms: unmethylated, hemimethylated or completely methylated. The dinucleotides cluster and are typically found in the promoter regions of genes, where methylation is generally associated with transcriptional silencing. Methylation can also be found outside of promoter regions, for example, in the gene body itself. Gene body methylation profiles are conserved, and in contrast to promoter methylation, most gene bodies are CpG-poor and extensively methylated. In contrast to promoter methylation, studies investigating the impact of methylation on gene bodies suggest that gene body methylation correlates with high gene expression in actively dividing cells. The function of gene body methylation is yet to be determined, but it is thought that it facilitates efficient transcription and represses harmful genetic elements such as transposable and viral elements.

Methylated CpGs are not equally distributed throughout the genome; most of the methylation is associated with gene bodies, repetitive sequences and intergenic regions. CpG islands are mostly unmethylated, especially when located in the promoter region of active genes. CpG islands are most commonly found in promoter regions of genes; approximately 60% of annotated genes have CpG island promoters. Only a small percentage of CpG island promoters responsible for tissue-specific gene expression and imprinting are methylated. The biggest variation in DNA methylation across cell types appears to happen in regions proximal to the CpG islands, typically within 2 kb that acquire disease- and tissue-specific methylation changes and are sometimes referred to as CpG shores. Conversely, exceptionally large low-methylated regions are termed methylated valleys or methylated canyons; they typically contain highly conserved, developmentally important genes that might be associated with cancer.

DNA methylation has a key role in regulating chromatin structure and function. DNA methylation is relatively more stable than histone post-transcriptional modifications; it is also reversible and can be removed through both active and passive mechanisms, both of which are vitally important for both cellular differentiation and normal development in mammals. Passive, for example during DNA replication, or active, for example, removed by specific enzymes (TET- 10-11 translocation), TDG (thymidine DNA glycosylase) or AID (activation-induced deaminase).

DNA methylation occurs via DNA methyl transferase enzymes (DNMTs) during the early stages of mammalian development and during the maturation of germ cells by DNMT3A and DNMT3B, assisted by the stimulatory factor DNMT3L. The methyl group is added to the carbon 5 position of cytosine through the action of the enzyme DNMTs. These enzymes (DNMT3A and DNMT3B) catalyse the de novo addition of a methyl group to cytosine in newly synthesised DNA. Only certain CpG sites are methylated, therefore the result is tissue- and cell-specific methylation patterns. These patterns are preserved once established, with only relatively small tissue-specific changes. Hypermethylation of CpG islands occurs in areas where there is a high density of CpG dinucleotides and is associated with gene silencing. Most CpG sites in the genome are methylated. CpG islands in the promoter regions of most human genes are not methylated, indicating that they are in a state of transcriptional activation (except for the case of imprinted genes that were silenced by DNA methylation).

During DNA replication, new unmethylated strands of DNA are made; these are then re-methylated by the maintenance methyltransferase DNMT1 – its role is as a methyl copy machine to ensure newly synthesised strands of DNA have the same methylation as the template strand for replication. This then enables DNA methylation to effect long-term transcriptional silencing, particularly of importance to repress repetitive elements, genomic imprints, X-chromosome inactivation, and to regulate specific gene expression during development and cellular specialisation. The multivalent interaction of DNMT3 enzymes with chromatin plays a key role in generating the genomic DNA methylation pattern. The cancer genome is typically characterised by global hypomethylation, which leads to genomic instability and focal hypermethylation, for example, of TS genes being silenced.

With the advent of large-scale NGS-based DNA methylation studies, the notion that DNA methylation occurs predominantly at CpG dinucleotides has been challenged. Recent studies comparing the DNA methylome in stem cells and foetal fibroblasts showed significant differences in terms of the composition and pattern of cytosine methylation between the two genomes. Approx. 25% of all cytosine methylation in stem cells was non-CG context compared to fibroblasts where almost all of the cytosine methylation was in the CG context. This suggests a role for DNA methylation in maintaining pluripotency in stem cells. The non-CG methylation was found to be enriched in gene bodies and depleted in protein binding sites and enhancers. Non-CG methylation in gene bodies was also positively correlated with gene expression.

Most tissue-specific DNA methylation occurs at alternative promoters found in gene bodies instead of 5′ promoters. The majority of CpG islands were found to be intragenic and intergenic regions. <3% CpG islands in 5′ promoters were methylated. In addition to traditional DNA methylome studies focusing on 5mC, there is now increasing data available on 5hmC. 5hmC is generated by TET proteins through oxidation of 5mC and is present at low levels in diverse cell types. It is assumed that 5hmC, like 5mC, has a role in transcriptional regulation, but it is likely that there is a difference between the two (Table 2.2).

2.3 DNA Demethylation

DNA demethylation is the process of removing methyl groups from DNA. This can occur through active or passive mechanisms. Active DNA demethylation involves the enzymatic removal of the methyl group by the Ten-Eleven

Table 2.2 DNA methylation.

DNA methylation location	Effect
Promoter region	Inhibits gene expression
Gene body region	Facilitates gene expression
Enhancer region	Inhibits gene expression
CpG island shores	Affects gene expression regulation
CpG island shelves	Affects gene expression regulation
Intergenic regions	Affects chromatin structure and gene expression
Repetitive DNA elements	Affects chromatin structure and gene expression

CpG islands are regions of DNA that contain a high density of CpG dinucleotides, while CpG island shores and shelves are regions immediately adjacent to CpG islands. Repetitive DNA elements include transposable elements and other repetitive sequences found throughout the genome.

Translocation (TET) family of enzymes, which oxidise the methyl group to a hydroxymethyl group that can be further processed and ultimately removed by DNA repair machinery. Passive DNA demethylation, on the other hand, involves the dilution of methylated DNA during DNA replication in the absence of DNMT activity, resulting in the progressive loss of methylation marks over time.

DNA demethylation can play a critical role in development and cellular differentiation, as it can allow for the activation of previously silenced genes. Moreover, aberrant DNA demethylation has been implicated in various diseases, including cancer and neurological disorders.

In developing mammals, demethylation is observed at two stages. The first occurs early on in embryogenesis in the paternal genome, just after fertilisation but before DNA replication. This first demethylation wave confers totipotency to the developing embryo. DNA methylation pattern is then re-established in the preimplantation stage of development. The second wave of demethylation occurs during germ cell specification, and this stage includes the specific demethylation of imprinted genes. Active demethylation also occurs at specific loci in T cells, neurons and other cells. The discovery of the role of TET enzymes in demethylation has increased our understanding of how DNA becomes demethylated through a stepwise oxidation process of 5mC to 5hmC, 5fC and finally to 5caC followed by removal of the higher oxidised bases by thymine DNA glycosylase and the base excision repair mechanism.

Removal of methylation allows the chromatin to revert to a less differentiated state, low levels of DNA methylation are needed for cellular reprogramming. Demethylation is a crucial step to ensure the erasure of the parental imprint; this can occur both passively and actively and has been observed in germline and in early embryogenesis.

There are three members in the TET family of enzymes: TET1, TET2 and TET3. Both TET1 and TET3 have a CXXC domain at their amino ends. It is thought that this domain is in part responsible for targeting the TET enzymes to CpG-containing regions of the genome; the CXXC domain has been shown to recruit DNMT1, MLL1 and CFP1 to unmethylated CpG sites. As the TET2 enzyme is without a CXXC domain, it is likely it relies on other proteins or transcription factors to recruit at specific loci. There is still much work to be done to elucidate the precise mechanisms through which TET enzymes act, what is known is that they do not function alone and interact with multiple proteins to modulate gene expression.

5mC is found at a consistent level of 4% of total cytosines across different somatic cell types. 5hmC is found at much lower levels and is more variable with respect to tissue type. It is most abundant in ES cells and the brain, where it is found between 0.4 and 0.7% of total cytosines and lower still in other tissues. Cancer cells often contain lower 5hmC than surrounding tissues. 5fC and 5caC are even less abundant, estimated to occur between 0.02–0.002% of total cytosines; however, the fact that they can be detected after several cell divisions suggest that they may be stable marks.

Active demethylation can occur in both dividing and non-dividing cells and usually occurs within six to eight hours of fertilisation prior to DNA replication. This occurs through either of two mechanisms that both convert hmC back to cytosine:

- Iterative oxidation by TET enzymes which continuously oxidise hmC
- Deamination by AID/APOBEC – activation-induced cytidine deaminase/apoliprotein B mRNA editing enzyme complex.

Passive demethylation occurs in dividing cells and is thought to occur in the maternal genome of mammals before pre-implantation growth. DNMT1 actively maintains DNA during replication, so simply excluding or down-regulating DNMT1 from the nucleus leaves any newly synthesised cytosines unmethylated; therefore overall methylation will be reduced after each round of division.

2.4   Chromatin Remodelling: Post-translational Modifications of Histone Proteins

Chromatin remodelling refers to the process of changing the structure and organisation of chromatin, the complex of DNA and histone proteins that make up chromosomes. Chromatin remodelling can involve the addition or removal of histone modifications, nucleosome repositioning and changes in the accessibility of DNA to transcription factors and other regulatory proteins. These changes can have a profound effect on gene expression by modulating the accessibility of DNA to the transcriptional machinery. Chromatin remodelling complexes are large, multi-subunit protein complexes that use the energy of ATP hydrolysis to alter the structure of chromatin. These complexes can be classified into two main groups: those that deposit histone modifications and those that move nucleosomes. Examples of chromatin remodelling complexes include the SWI/SNF family of complexes, the ISWI family of complexes and the Polycomb group (PcG) of proteins.

Chromatin exists in two forms: a condensed form during cell division known as heterochromatin and a more relaxed form, euchromatin, which provides the environment for DNA regulatory activity. During cell division, chromosomal DNA is heavily compacted to enable the segregation of sister chromatids into daughter cells, for the rest of the time, each chromosome exists in a distinct location of the nucleoplasm. Chromosomal DNA is packaged through nucleosomes; discs ~ 11 nm in diameter and 5.5 nm in height comprising an octet of histone proteins ($H3_2$-H42) (H2A-H2B)$_2$ and 147 bp of DNA. The amount of DNA linking two nucleosomes ranges from less than 10 bp to well over 90 bp. Nucleosomes stack in solenoidal structures. The chromatin needs to be relaxed for the DNA regulatory process to function; chromatin modification occurs through the addition of covalent molecules to the histones. Methylation of histone H3 lysine 4 (H3K4) and H3 lysine 36 is associated with transcription activation; conversely, methylation of H3 lysine 9, H3 lysine 27 and H4 lysine 20 is correlated with transcriptional repression.

The histones have protruding N-terminal tails; it is these that undergo post-translational modifications, including acetylation, methylation, phosphorylation, ubiquitination and SUMOylation. In terms of nomenclature, the histone is named, followed by the modified amino acid residue and its position in the protein and then the type of modification; H3K27me3 refers to 3 methylation groups on lysine 27 in the histone H3 tail. Arginine (R) and lysine (L) are both frequently found to be acetylated or methylated. Histone modifications are reversible and have a role in either repressing or enabling active chromatin, e.g., H3K9me2/3 and H3K27me3, both gene repression, H3K4me3 and H3/H4 acetylation associated with gene expression (Table 2.3).

It is possible for a histone to be a target of both methylation and acetylation, such as in H3K9 and H3K27; these are mutually exclusive events so need to step to reverse from repressive to active or vice versa. H3K9m3 and H3K27me3 are associated with gene repression, H3K4me1 and H3K4me3 and acetylation with activation. Histone modifications are another important epigenetic mechanism in transcriptional regulation. The nucleosome is comprised of four core histones around which is wrapped 146 bp of DNA. The N-terminal tails of histone polypeptides can be modified by over 100 different post-translational modifications; these include methylation, acetylation, phosphorylation and ubiquitination – collectively these are referred to as histone modifications and are not as well studied or understood as DNA methylation. As in the case of DNA methylation, histone

Table 2.3 Histone modifications.

Histone modification	Location	Function
H3K4me3	Promoters	Active transcription initiation
H3K4me1	Enhancers	Enhance transcription initiation
H3K27ac	Enhancers, Promoters	Enhance transcription
H3K27me3	Gene bodies, Promoters	Transcriptional repression
H3K36me3	Gene bodies	Facilitates transcriptional elongation
H3K9me3	Heterochromatin	Transcriptional repression
H2BK120ub	Gene bodies	Facilitates transcriptional elongation
H3K79me2	Gene bodies	Facilitates transcriptional elongation
H4K16ac	Chromatin boundaries	Condensation and folding of chromatin

modifications can regulate transcription through either modifying the structure of the chromatin or through bringing about condensing of the chromatin.

There are at least eight distinct types of histone modification including acetylation, methylation, phosphorylation, ubiquitination, sumoylation, ADP ribosylation, deamination and proline isomerisation. It is believed these different marks affect the nucleosome-nucleosome or DNA-nucleosome interactions and represent docking sites for the recruitment of specific proteins that can result in different cellular outcomes. Added complexity is that it is likely that these modifications are interdependent and work in specific combinations, meaning different modifications can result in distinct and consistent cellular outcomes; 'histone code'. These modifications and others can act in combination to create a complex 'histone code' that regulates gene expression and other chromatin-based processes. The specific effects of a given histone modification can vary depending on the context, including the location of the modification on the histone protein, the identity of neighbouring histone modifications and the presence of other chromatin-associated proteins (Table 2.4).

Proline isomerisation plays a significant role in protein folding. Proline can occupy both isomers (cis and trans) when forming a peptide bond with other amino acids to regulate gene expression. Histones 2A and 2B have multiple proline residues that will affect the activity of the histone. The isomerisation of the peptide bond between histone H3's alanine 15 and proline 16 is affected by the acetylation at K14, promoting the trans isomer of P16, which in turn reduces K4 methylation repressing gene transcription. Frp4 is a histone isomerase of prolines 30 and 38 on the histone H3 tail. The confirmation of P38 is necessary for the induction of lysine 36 of histone H3 (H3K36) methylation; its isomerisation appears to inhibit the ability of Set2 to methylate H3K36.

Table 2.4 Table of some of the common histone modifications and their associated effects.

Histone modification	Abbreviation	Associated effect
Acetylation	Ac	Neutralises the positive charge of histones, loosening the chromatin structure and promoting gene expression
Methylation	Me	Can either promote or repress gene expression, depending on the location and degree of methylation
Phosphorylation	Ph	Often associated with changes in chromatin structure and gene expression, particularly during cell cycle regulation and stress response
Ubiquitination	Ub	Can either promote or repress gene expression, depending on the location and degree of ubiquitination
Sumoylation	Su	Like ubiquitination, can either promote or repress gene expression, depending on the location and degree of sumoylation

Sumoylation is the addition of a 'small ubiquitin-related modifier protein' (SUMO). The addition of the ~100 amino acid protein has a role in transcription repression by opposing other active marks such as acetylation and ubiquitination, potentially through the recruitment of HDACs and HP1 proteins. Putative sumoylation sites are K6/7, K16/17 of H2B and K126 of H2A, all four lysines of the N-terminal tail of H4.

Ubiquitin is a highly conserved 76 amino acid protein. Ubiquitination involves the addition of ubiquitin to a lysine residue. Typically, the attachment of one ubiquitin modifies protein function, whereas the addition of multiple ubiquitin (polyubiquination) marks a protein to be degraded via the 26S proteasome. Histone H2A ubiquitination is dependent on the polycomb repressive complex 1 (PRC1); PRC2 sets up the H3K27me3 marks that are recognised by the PRC1 complex, which would ubiquitin H2A and silence gene expression, ubiquitination of H2A at K119 has been shown to be important for transcriptional activation. Similarly, ubiquitination of H2B has been linked to both transcriptional activation and inhibition. Ubiquitination is likely to affect other histone modifications; H3K4 and H3K79 methylation were shown to be dependent on Rad6-mediated H2BK123 ubiquitination. The ability of ubiquitination to impact upon methylation makes sense of the role of ubiquitination in both activation and inhibition of transcription.

Histone ADP-ribosylation involves the addition of an ADP-ribose onto a protein using NAD+ as a substrate. Mono-ADP-ribosylation on H4 appears to occur preferentially when H4 is hyperacetylated, and mono-ADP-ribosylation of histone H1.3 on arginine 33 may reduce cyclic AMP-dependent phosphorylation of serine 36.

Histone phosphorylation is the addition of a phosphate (PO4) group to a protein molecule, catalysed by protein kinases; phosphatases remove the phosphate group. The most studied histone phosphorylation is serine 10 of histone H3 (H3S10). Its role in gene expression possibly is linked to chromatin condensing and chromosome segregation; research has linked the mark to transcriptional activation and inhibition. Also believed to have a role in DNA damage repair, phosphorylation of H2AX and in recruitment of DNA repair factors at DNA break sites.

Histone methylation is the addition of a methyl group in histones only to arginine and lysine, arginine can be mono or dimethylated, and lysines can be mono, di or trimethylated. Methylated marks on histones could be related to activation, elongation, or repression of gene expression. H3K9, H3K27 and H4 K20 are all linked to transcriptional repression. Histone methylation also has a role in DNA damage response and DNA repair.

Histone acetylation transfers an acetyl group from acetyl co-a to the lysine epsilon amino group on the n-terminal tails of histones catalysed by histone acetyltransferases, this can occur on H3, H4, H2B and H2A. Hyperacetylation of histones is associated with transcriptionally active regions but can also affect other cellular processes, including DNA repair and replication.

Nucleosomal histones can recruit proteins that have histone-binding domains. The promotion or inhibition of this interaction can impact the function of the underlying DNA. Such a disruption typically occurs through the modification of lysine and arginine residues on the histone. Terminology relating to how this regulation is interpreted typically uses the terms 'readers' 'writers' and 'erasers' that interpret, deposit, or remove modifications to the histone residues, respectively. Writers typically refer to enzymes such as methylases, acetylases and ubiquitylases and erasers such as demethylases, deacetlyases and deubiquitylases. Overall, the specific histone modifications involved in epigenetic gene control can vary depending on the cell type, developmental stage and environmental cues, highlighting the dynamic nature of epigenetic regulation.

2.5 Non-coding RNAs

ncRNAs are RNA molecules that do not code for proteins but instead have a regulatory role in various cellular processes, including epigenetic gene control. There are several types of ncRNAs, including long non-coding RNAs (lncRNAs), microRNAs (miRNAs) and small interfering RNAs (siRNAs), among others. These ncRNAs can act through a variety of mechanisms to modulate gene expression, including by regulating chromatin structure, DNA methylation and histone

Table 2.5 Table summarising some of the several types of non-coding RNAs (ncRNAs) and their functions.

Type of ncRNA	Length	Function
microRNA (miRNA)	~22 nucleotides	Regulation of gene expression through binding to complementary mRNA, leading to degradation or translational repression
small interfering RNA (siRNA)	~20–25 nucleotides	Like miRNA, but derived from double-stranded RNA molecules, often induced by viral infections
long non-coding RNA (lncRNA)	>200 nucleotides	Regulation of gene expression through diverse mechanisms, such as chromatin modification, transcriptional regulation and mRNA stability
circular RNA (circRNA)	Circular	Regulation of gene expression through a variety of mechanisms, including interaction with miRNAs and other RNA-binding proteins
piwi-interacting RNA (piRNA)	24–31 nucleotides	Regulation of transposable elements and maintenance of genome stability in germ cells
ribosomal RNA (rRNA)	>100 nucleotides	Structural component of ribosomes, which synthesise proteins
transfer RNA (tRNA)	~70–90 nucleotides	Transports amino acids to the ribosome for protein synthesis
small nucleolar RNA (snoRNA)	~60–300 nucleotides	Involved in the processing and modification of rRNA

modifications. ncRNAs are also believed to be important regulators of gene expression with a key role in development and differentiation. They are a collective term for several distinct types of ncRNA; including short interfering RNAs – siRNA, microRNAs – miRNA, PIWI-interacting RNAs – piRNAs and long ncRNAs – lncRNAs which are longer than 200 nt. ncRNAs act through histone modification and through the recruitment of DNMTs (Table 2.5).

2.6 Polycomb Proteins

Polycomb proteins are a group of proteins that play a critical role in epigenetic gene regulation, particularly during development and differentiation. These proteins are part of a conserved complex that modifies histones and influences chromatin structure, leading to the repression of gene expression. Polycomb proteins act by forming repressive complexes that modify histones, typically through the addition of methyl groups to lysine residues on histone tails. These modifications, known as H3K27me3 and H2AK119ub, create a chromatin environment that is less accessible to transcription factors and RNA polymerase, leading to the repression of gene expression. Polycomb proteins are involved in the regulation of a wide range of developmental genes, including those involved in cell fate determination and differentiation. They play a critical role in maintaining the identity of stem cells and preventing their differentiation prematurely. Additionally, recent studies have suggested that polycomb proteins may also play a role in the regulation of ncRNA genes.

Polycomb complexes are found in the nuclei of most cells and have a significant role in gene expression. They specifically target genes in major differentiation pathways and those that are responsible for governing the development and identity fate of a cell. They also integrate signals from other cells, growth factors, morphogens or other external stimuli that contribute to a cell's response to its environment. The Polycomb complex comprises two multiprotein complexes: PRC1 and PRC2. The two complexes generate two types of histone modification: H2A ubiquitylation and H3K27 methylation. The PRC2 complex is responsible for all H3K27 methylation. The H3K27me3 signature is highly correlated with silent loci and is important for regulating developmental and oncogenic genes. In any given cell, some polycomb targets bind polycomb proteins and are repressed, whereas others remain active. Polycomb response elements are genomic regions spanning several hundred base pairs that can recruit both PRC1 and PRC2 complexes, resulting in polycomb repression of neighbouring genes. In drosophila, PRE can repress several genes, often remotely from their promoters (up to tens of Kb away). PRC2 can acquire both gain-of-function and loss-of-function mutations that are associated with aggressive cancers (Comet et al. 2016). However, the intracellular determinants of PRC2 chromatin binding are not well understood.

This is maintained by epigenetic maintenance/cellular memory. That is, once the polycomb proteins establish repression, this state is remembered in successive cell cycles and re-established in cell progeny. Similarly, if the gene was active in the early embryo, this would prevent polycomb repression at a later stage. Polycomb repression is easily reversed by the presence of activators, so in terms of an epigenetic mark, polycomb mechanisms can be viewed as a mechanism that raises the threshold of signals or activators to turn on a gene; the higher threshold is also then transmitted to cellular progeny, and as a result, these mechanisms can modify the ability of cells to respond to transcriptional signals depending on the history of the cell lineage. This is important, particularly in the development of pattern formation, morphogenesis and organogenesis. In mammals, this appears to be done through DNA methylation and not polycomb mechanisms. In mammals, PRC2 accessory proteins are often found bound to polycomb target genes; their depletion results in a reduction of PRC2 and H3K27me3 at some, but not all, of the PRC2 target genes. Polycomb mechanisms are likely to respond to metabolic stress, intracellular signalling, stress response and many other factors in development, again, homeostasis and disease are interesting areas to study in the context of human health and disease.

Polycomb protein	Function in epigenetic control
EZH2	Methylates histone H3 at lysine 27, resulting in transcriptional repression of target genes. EZH2 is frequently overexpressed in cancer, leading to abnormal gene silencing.
EED	Forms a complex with EZH2 and other proteins to establish and maintain H3K27me3 marks, contributing to gene repression.
SUZ12	Forms a complex with EZH2 and EED to silence gene expression through the establishment of H3K27me3 marks. It also plays a role in regulating the activity of EZH2.
RING1A and RING1B	Methylate histone H2A at lysine 119, which works together with H3K27me3 to repress target genes. The RING1A/RING1B complex is part of the Polycomb Repressive Complex 1 (PRC1).
CBX proteins	Bind to H3K27me3 and/or H2AK119ub1 marks to recruit PRC1 to target genes, resulting in transcriptional repression. Different CBX proteins can contribute to tissue-specific gene repression.
BMI1	Part of PRC1 and required for its activity. It has been implicated in the maintenance of stem cell self-renewal and the development of cancer.
PHC proteins	Part of PRC1 contributes to H3K27me3 recognition and recruitment of PRC1 to target genes. They also interact with other chromatin-modifying enzymes and transcription factors.

2.7 Conclusion

Epigenetics is the study of heritable changes in gene expression or cellular phenotype that occur without alterations to the underlying DNA sequence. In other words, epigenetic changes can affect how genes are expressed without changing the actual genetic code. These changes can be influenced by a variety of factors, such as environmental exposures, lifestyle choices and age. Epigenetic modifications can include DNA methylation, histone modifications and ncRNA molecules, and they can have important roles in development, cellular differentiation and disease susceptibility. Overall, DNA methylation is a critical epigenetic mechanism for regulating gene expression in development, ageing and disease, and dysregulation of DNA methylation has been implicated in numerous human disorders. Recent research has highlighted the dynamic interplay between DNA methylation and demethylation, and the complex regulatory mechanisms that control these processes. A deeper understanding of the mechanisms of DNA demethylation may lead to new therapeutic approaches for the treatment of human diseases. The role of chromatin remodelling in epigenetic gene control is critical, as the accessibility of DNA to the transcriptional machinery is a key determinant of gene expression. Aberrant chromatin remodelling has been implicated in various human diseases, including cancer and neurological disorders. Therefore, understanding the mechanisms of chromatin remodelling and their regulation is an active area of research in epigenetics and biomedical science. In addition to their role in gene regulation, ncRNAs have also been implicated in various human diseases, including cancer and neurological disorders, and they represent a promising target for the development of novel therapies. Overall, the study of ncRNAs and their role in epigenetic gene control is an active and rapidly evolving area of research in molecular biology and biomedical science. Dysregulation of Polycomb proteins has been implicated in various human diseases, including cancer and developmental disorders. For example, aberrant Polycomb-mediated gene repression can lead to the silencing of tumour suppressor genes, contributing to the development of cancer. On the other hand, abnormal activation of Polycomb complexes can result in the repression of genes involved in differentiation, leading to developmental disorders.

Task

- *What is known about gene regulation for your topic?*
- *Is your proposal likely to be due to epigenetic/epigenomic effects and fixed/responsive?*
- *What is your population; what are your phenotypes?*

Reference

Comet I, Riising EM, Leblanc B, Helin K. Maintaining cell identity: PRC2-mediated regulation of transcription and cancer. *Nat Rev Cancer*. 2016 Dec;16(12):803–10. doi: 10.1038/nrc.2016.83. Epub 2016 Sep 23. PMID: 27658528.

The article discusses the role of PcG proteins and their enzymatic complex PRC2 in maintaining cellular identity through epigenetic regulation of gene expression. It provides an overview of the molecular mechanisms of PRC2-mediated gene silencing and its involvement in normal development and cancer progression. The authors also discuss recent advances in understanding the regulation of PRC2 activity and potential therapeutic strategies targeting this complex in cancer treatment.

Further Reading

Capell BC, Berger SL. Genome-wide epigenetics. *J Invest Dermatol.* 2013 Jun;133(6):e9. doi: 10.1038/jid.2013.173. PMID: 23673507; PMCID: PMC4824393.

This article provides an overview of the field of genome-wide epigenetics, which studies the epigenetic modifications occurring on a genome-wide scale. The authors describe the different epigenetic marks, such as DNA methylation and histone modifications, and their impact on gene expression and cellular function. They also discuss the different technologies used to study genome-wide epigenetics, such as ChIP-seq, bisulphite sequencing and whole-genome shotgun bisulphite sequencing, and the challenges associated with analysing and interpreting large-scale epigenetic data sets. The authors emphasise the importance of genome-wide epigenetic studies in advancing our understanding of complex diseases and the potential of epigenetic therapies for personalised medicine.

Flanagan JM. Epigenome-wide association studies (EWAS): past, present, and future. *Methods Mol Biol.* 2015; 1238:51–63. doi: 10.1007/978-1-4939-1804-1_3. PMID: 25421654.

The article provides an overview of the history, current state and future of epigenome-wide association studies (EWAS). The author discusses the development of EWAS, which involve the systematic analysis of epigenetic modifications across the entire genome to identify associations with disease, environmental factors and other phenotypes. The article covers the challenges and limitations of EWAS, including issues related to sample size, data analysis and replication. The author also highlights recent advances in technology and methods that have improved the accuracy and resolution of EWAS. The article concludes with a discussion of future directions for EWAS research, including the potential for integrative analyses with genomic and transcriptomic data and the use of EWAS in personalised medicine. Overall, the article provides a valuable resource for researchers interested in the application of epigenetics to the study of complex diseases and other phenotypes.

Gluckman PD, Hanson MA, Low FM. The role of developmental plasticity and epigenetics in human health. *Birth Defects Res C Embryo Today.* 2011 Mar;93(1):12–8. doi: 10.1002/bdrc.20198. PMID: 21425438.

This article discusses the significant role of developmental plasticity and epigenetics in human health. It highlights how environmental cues during early development can influence long-term health outcomes and how these effects can be mediated by epigenetic mechanisms. The authors provide examples of how nutritional, hormonal and other environmental factors can alter gene expression and contribute to disease risk later in life. They also discuss the potential for epigenetic interventions to prevent or treat such diseases. The article emphasises the importance of understanding the complex interactions between genetics, environment and epigenetics in shaping human health and disease.

Inbar-Feigenberg M, Choufani S, Butcher DT, Roifman M, Weksberg R. Basic concepts of epigenetics. *Fertil Steril.* 2013 Mar 1;99(3):607–15. doi: 10.1016/j. fertnstert.2013.01.117. Epub 2013 Jan 26. PMID: 23357459.

This is a comprehensive review of the basic concepts of epigenetics. They cover the different types of epigenetic modifications, including DNA methylation, histone modifications and ncRNAs. The authors also discuss the mechanisms of epigenetic regulation and their role in development, differentiation and disease. They further describe the methods used for studying epigenetic modifications and their potential use as biomarkers for the diagnosis and prognosis of diseases.

Kalish JM, Jiang C, Bartolomei MS. Epigenetics and imprinting in human disease. *Int J Dev Biol.* 2014;58(2–4):291–8. doi: 10.1387/ijdb.140077mb. PMID: 25023695.

This paper focuses on the role of epigenetics and imprinting in human disease. The authors discuss the mechanisms of imprinting and epigenetic regulation, including DNA methylation and histone modification, and their role in normal development and disease. The article also reviews several human diseases associated with epigenetic alterations, including cancer, imprinting disorders and neurological and metabolic disorders. The authors highlight the importance of understanding the epigenetic mechanisms underlying these diseases for diagnosis and treatment and discuss the potential for epigenetic therapies in the future. The article concludes by emphasising the need for continued research to fully understand the role of epigenetics in human disease.

Skinner MK. Role of epigenetics in developmental biology and transgenerational inheritance. *Birth Defects Res C Embryo Today.* 2011 Mar;93(1):51–5. doi: 10.1002/ bdrc.20199. PMID: 21425441; PMCID: PMC5703206.

This paper discusses the importance of epigenetics in developmental biology and transgenerational inheritance. It provides an overview of epigenetics and its role in gene expression, development and inheritance. The potential for environmental factors to affect epigenetic marks is also discussed, leading to changes in gene expression that can be passed down to future generations. The paper highlights the importance of studying epigenetic mechanisms to better understand how environmental factors can influence gene expression and disease susceptibility across generations. Skinner's article provides a comprehensive introduction to epigenetics and its potential implications in developmental biology and transgenerational inheritance.

3

Epigenetic Mechanisms, Homeostasis and Potential for Manipulating the Epigenome

Having now attained an understanding of how epigenetics works, we will explore further whether it is feasible, for a given condition, that the epigenome can be manipulated, and if, so how. So, the aim here is, if you are going to do research looking at epigenomic biomarkers, what is it that you hope to achieve in terms of a potential translational effect for human health? We will focus on two broad areas: pharmacological and environmental intervention. Pharmacological intervention will look at how we can intervene with the molecular machinery controlling gene regulation, both through pharmaceuticals and potentially through nutrient intervention. Environmental will look at the contribution of environment, stress, and lifestyle and how these could be modified to manipulate a 'maladapted' epigenome found in a particular disease state. We will also focus on the inflammatory response and the role of epigenetic reprogramming of the cells of the immune system. The chapter will end by considering transgenerational epigenetic inheritance; this is an important consideration, especially if interventions are targeted at manipulating the epigenome.

3.1 Pharmaceutical

Pharmaceutical manipulation of the epigenome involves the use of drugs to modify epigenetic marks, such as DNA methylation and histone modifications, to alter gene expression patterns. This approach is being actively pursued as a potential therapeutic strategy for a wide range of human diseases, including cancer, neurodegenerative disorders and cardiovascular diseases. Drugs that target epigenetic mechanisms include methylation-inhibiting drugs, bromodomain inhibitors, histone acetylase (HAT) inhibitors, protein methyltransferase (PMT) inhibitors, histone methylation inhibitors and histone deacetylase (HDAC) inhibitors. Methylation-inhibiting drugs are the oldest epigenetic drugs and have been around

Epigenetics and Health: A Practical Guide, First Edition. Michelle McCulley.
© 2024 John Wiley & Sons, Inc. Published 2024 by John Wiley & Sons, Inc.

for more than 50 years, although their mechanism of action has only recently become elucidated. These include nucleoside-like compounds, several of which have been approved for treating cancer, including 5-Azacytidine and Zebularine. They both work similarly to produce cytotoxic effects; the drug, once in the cell, becomes phosphorylated and incorporated into the DNA during replication, this is recognised by, and forms a covalent bond with, DNMT1, which then triggers degradation of the enzyme and widespread reduction in methylation. Other strategies include the development of antisense oligonucleotides and other small molecules designed to decrease levels of DNMT1 or directly inhibit the DNMT1 active site.

Bromodomain inhibitors have been found to be effective in down-regulating c-Myc through the disruption of the interaction between bromodomain and extra-terminal (BET) proteins and acetylated histones. JQ1 is a selective bromodomain inhibitor that has demonstrated widespread downregulation of c-Myc in murine models of multiple myeloma. I-BET726 has been found to inhibit tumour growth and is highly selective for the BET family proteins. It also has a direct regulatory effect on the expression of the antiapoptotic gene, *BCL2* that is found to be highly expressed in several tumours. HAT inhibitors target the catalytic activity of HATs in many cancers and other diseases. They are not particularly selective and will bind a range of proteins; however, they are viewed as promising pharmaceuticals for a range of diseases. PMT inhibitors target the methylation of lysine and arginine residues, impacting gene transcription. BIX-01294 was the first selective inhibitor of the protein lysine methyltransferase. A potent, selective inhibitor of PMT with low cell toxicity is UNC0638.

Histone methylation inhibitors inhibit trimethylation and reactivate developmentally regulated genes. DZNep has been reported to reactivate silenced genes in cancer cells as well as selectively inhibit the trimethylation of lysine 27 on histone H3 (H3K27me3) and lysine 20 on histone H4 (H4K20me3). HDAC inhibitors are diverse structurally and can be categorised based on their structure, evidence suggests that HDACi only affect between 2 and 10% of expressed genes. The drug Vorinostat results in hyperacetylation of histones and non-histone proteins such as p53 and HSP90 that then induce apoptosis and sensitise tumours to cell death processes and other pharmaceuticals. Unfortunately, the negative consequence of Vorinostat having multiple targets is that it also results in many side effects (Table 3.1).

3.2 DNA Methylation and Impact on Nutrition

Emerging evidence suggests that nutrition can modulate DNA methylation patterns, leading to changes in gene expression that may have implications for human health. One example of the impact of nutrition on DNA methylation is in the

Table 3.1 Overview of methods used to manipulate the epigenome.

Method*	Description	Example
DNA methylation inhibitors	Drugs that block DNA methyltransferases, leading to global or targeted DNA demethylation	5-aza-2′-deoxycytidine (decitabine)
Histone deacetylase inhibitors	Drugs that inhibit histone deacetylases, leading to increased acetylation of histones and gene expression	Vorinostat (Zolinza)
Histone methyltransferase inhibitors	Drugs that inhibit histone methyltransferases, leading to decreased methylation of histones and potential changes in gene expression	BIX-01294
Histone demethylase inhibitors	Drugs that inhibit histone demethylases, leading to increased methylation of histones and potential changes in gene expression	GSK-J4
CRISPR-Cas9	Gene editing technology that can be used to introduce specific epigenetic modifications at targeted loci	Targeted DNA methylation, targeted histone modification, etc.
Epigenetic therapies	A range of approaches, including small molecules and biologicals, aimed at modulating the epigenome to treat diseases, such as cancer and neurological disorders	Epacadostat, Entinostat

* Note that these are just a few examples of the many methods that are currently being used or developed to manipulate the epigenome. The specific approach used will depend on the desired target and the specific details of the epigenetic modifications involved.

context of foetal development. Maternal nutrition during pregnancy can influence DNA methylation patterns in the developing foetus, which can have lasting effects on health outcomes in adulthood. For example, maternal folate deficiency has been associated with altered DNA methylation patterns in the offspring, leading to an increased risk of obesity, type 2 diabetes and other chronic diseases. Nutrition can also impact DNA methylation patterns in adults. Studies have shown that diets rich in methyl donors, such as folate, methionine and choline, can increase DNA methylation levels, whereas diets deficient in these nutrients can lead to DNA hypomethylation. Additionally, certain dietary components, such as polyphenols and flavonoids found in fruits and vegetables, have been shown to influence DNA methylation patterns and gene expression.

There is now evidence linking early-life nutritional deprivation with stable DNA methylation changes and increased risk of chronic disease later in life. Key examples are often cited as evidence including the Dutch famine in the Second World War and the Chinese great famine (1959–61). The Dutch famine study identified significantly differentiated and persisting methylation of *IGF2* (insulin growth factor 2) in response to the famine in utero, and follow-up studies on the cohort linked these methylation changes with adult metabolic disease, T2DM and schizophrenia. The Chinese great famine resulted in approximately 30 million deaths, the offspring of mothers who survived the famine went on to be shorter with worse midlife health and higher rates of chronic diseases, including schizophrenia. A recent study by He et al. (2019) provided experimental evidence that during the biological adaptation to famine conditions, DNA methylation changes occur, which predominantly affect genes involved in the central nervous system. In a recent study, DHA supplementation in pregnancy could modulate some of the adverse effects of maternal obesity by influencing *IGF2* methylation. The impact of nutrition on DNA methylation has important implications for human health and disease. Aberrant DNA methylation patterns have been linked to a wide range of diseases, including cancer, cardiovascular disease and neurodegenerative disorders. By modulating DNA methylation patterns through nutrition, it may be possible to prevent or treat these diseases, although much work remains to be done to understand the mechanisms underlying these effects and to identify safe and effective dietary interventions.

3.3 Diet and Cancer Prevention

Disruptions in epigenetic regulation are key drivers in cancer progression. Due to the potential reversibility of epigenetic marks, they are good potential therapeutic targets. Many drugs are designed to target epigenetic changes, but there is also research looking into the chemo-preventative properties of nutrition and potential dietary epigenetic regulators (Table 3.2).

Bioactive components of our diet can act as epigenetic regulators by influencing DNA methylation, histone modifications and ncRNA expression and function. Chemoprevention can be described as primary such as avoiding carcinogen exposure; secondary, such as blocking, slowing or reversing cancer progression or tertiary where cancerous lesions are subdued or removed. It is suggested that nutritional bioactive compounds can be effective at all these stages, albeit complicated due to the many unique bioactive metabolites produced upon the metabolism of nutritional products. There is also the issue of threshold, which is the metabolites must be of sufficient concentration when they enter the circulation so that they can reach the target tissue. So how effective a dietary component will be in terms of chemoprevention depends on the bioavailability of the bioactive compound; this will also be affected by the individual's epigenetic,

Table 3.2 Impact of nutrients on DNA methylation.

Nutrient/compound	Effect on DNA methylation	Food sources
Folate	Increases DNA methylation	Leafy green vegetables, legumes, fortified cereals
Vitamin B12	Increases DNA methylation	Meat, fish, dairy, fortified cereals
Choline	Increases DNA methylation	Eggs, liver, beef, wheat germ
Betaine	Increases DNA methylation	Beets, spinach, whole grains
Methionine	Decreases DNA methylation	Meat, fish, dairy, eggs, nuts, seeds
S-adenosylmethionine (SAM)	Increases DNA methylation	Meat, fish, dairy, legumes, whole grains
Flavonoids	Increases or decreases DNA methylation (depending on the type and dose)	Fruits, vegetables, tea, cocoa
Resveratrol	Increases DNA methylation	Grapes, wine, berries, peanuts
Curcumin	Increases DNA methylation	Turmeric, curry
Sulforaphane	Increases DNA methylation	Cruciferous vegetables (broccoli, cauliflower, kale, etc.)
Green tea catechins	Increases DNA methylation	Green tea
Omega-3 fatty acids	Increases or decreases DNA methylation (depending on the dose)	Fatty fish, fish oil supplements, flaxseeds, chia seeds

genetic and environmental effects. Ideal chemo-preventative components would work at the initiation phase of cancer progression to prevent the onset of the disease (Table 3.3).

3.4 Dietary DNMT Inhibitors

Diet can influence the activity of the DNMT enzymes. Promoter hypermethylation of tumour suppressor genes is commonly found in cancers, so DNMT inhibitors are explored as pharmaceutical targets for epigenetic therapy. Azacytidine and decitabine are FDA-approved synthetic DNMT inhibitors used to treat myelodysplastic syndrome and acute myeloid leukaemia. Because of their pleiotropic

Table 3.3 Potential chemo-preventative nutrients and their proposed mechanism of action.

Epigenetic mechanism	Nutrient	Food source	Cancer type	Proposed mechanism
DNA methylation	Folate	Dark green leafy vegetables, legumes, citrus fruits, whole grains	Colorectal	Donor for the methylation of DNA and regulates gene expression
DNA methylation	Selenium	Brazil nuts, seafood, meat, eggs	Prostate	Promotes DNA methylation and protects against oxidative stress
Histone modification	Curcumin	Turmeric	Various	Inhibits histone deacetylases and activates histone acetyltransferases
Histone modification	Resveratrol	Grapes, wine, berries, peanuts	Breast, Prostate	Activates histone acetyltransferases and inhibits histone deacetylases
Non-coding RNA	Soy Isoflavones	Soybeans, tofu, miso	Breast	Targets non-coding RNA and regulates gene expression
Non-coding RNA	Epigallocatechin-3-gallate (EGCG)	Green tea	Various	Targets non-coding RNA and regulates gene expression

molecular effects and systemic toxicity, these drugs cannot be used as a primary preventative strategy in healthy people. Hence the attraction of identifying less noxious, dietary-derived DNMT inhibitors.

Dietary polyphenols are an example of DNMT inhibitors; these include genistein (from soy) and EGCG (from green tea). There is insufficient evidence in the literature to date supporting the efficacy of these bioactive compounds, and no studies looking at their impact on epigenetic marks so it is difficult at present to make conclusions as to whether they are effective.

Bioavailability, dosing, timing of exposure to the agent and the existing individual methylation pattern may all influence a person's response to a bioactive food component and need to be assessed within the context of the entire diet and of the individual. Cancer also has an exceptionally long latency period, so it is extremely hard to know the precise time point for an ideal dietary intervention (Table 3.4).

Table 3.4 Food compounds reported to inhibit DNMT.

Food/ compound	Description	Evidence for DNMT inhibition
Green tea	A type of tea made from Camellia sinensis leaves	Epigenetic changes associated with green tea polyphenols include DNMT inhibition
Curcumin	A compound found in turmeric root	Inhibits DNMT1 expression and activity in multiple cancer cell lines and animal models
Sulforaphane	A compound found in cruciferous vegetables such as broccoli, Brussels sprouts and kale	Inhibits DNMT activity in cancer cell lines and animal models
Resveratrol	A compound found in grapes, red wine and some berries	Inhibits DNMT activity and induces global hypomethylation in cancer cell lines and animal models
Genistein	An isoflavone found in soybeans and soy products	Inhibits DNMT activity and reactivates tumour suppressor genes in cancer cell lines and animal models
Quercetin	A flavonoid found in many fruits and vegetables	Inhibits DNMT activity and induces hypomethylation in cancer cell lines and animal models
Folate	A B vitamin found in many foods, including leafy green vegetables, fruits and beans	Acts as a cofactor for DNMTs, but high doses may have paradoxical effects and promote cancer development
Selenium	A mineral found in many foods, including nuts, seafood and meats	May inhibit DNMT activity in cancer cells, but the evidence is limited and conflicting

3.5 Dietary HDAC Inhibitors

Four HDAC inhibitors are FDA-approved for the treatment of lymphoma and multiple myeloma, but like DNMT inhibitors, their effect is pleiotropic in terms of their impact on gene expression. Allyl derivatives from garlic have been shown to induce histone acetylation in various human cancer cells. Allyl mercaptan has been shown to hyperacetylate *CDKN1A*, increase *CDKN1A* gene expression and promote cell cycle arrest. Raw garlic has been shown to influence the expression of multiple immunity- and cancer-related genes. In humans, increased consumption of cruciferous vegetables has been associated with decreased risk for cancer development, potentially via HDAC inhibition. Although in infancy and with many issues to take on board, HDAC inhibition is still promising for chemoprevention, by targeting histones, the chromatin structure is influenced and affects gene expression at many levels, meaning HDAC inhibitors can affect a diverse range of cellular functions. However, a much better understanding of the range of effects and improvement of bioavailability needs to be addressed before therapeutic efficacy can be achieved (Table 3.5).

Table 3.5 Food sources reported to impact chemoprevention via HDAC inhibition.

Food/sources	Compounds	Notes
Garlic	Allicin, diallyl disulfide, S-allylmercaptocysteine	Garlic and its derivatives are potent HDAC inhibitors. They also have anti-inflammatory and antioxidant properties.
Onions	Quercetin, fisetin	Onions are rich in quercetin, a flavonoid compound that has HDAC inhibitory activity. Fisetin is another flavonoid found in onions that has been shown to inhibit HDACs.
Turmeric	Curcumin	Curcumin is the active ingredient in turmeric and is a potent HDAC inhibitor. It has been shown to have anti-inflammatory, antioxidant and anticancer properties.
Green tea	Epigallocatechin-3-gallate (EGCG)	EGCG is a major component of green tea and has been shown to have HDAC inhibitory activity. Green tea also has antioxidant properties and has been associated with cancer prevention.
Grapes	Resveratrol	Resveratrol is a polyphenolic compound found in grapes and has HDAC inhibitory activity. It has been shown to have antioxidant and anti-inflammatory properties and has been associated with cancer prevention.
Dark chocolate	Epicatechin	Epicatechin is a flavonoid compound found in dark chocolate that has HDAC inhibitory activity. Dark chocolate has also been shown to have antioxidant properties and has been associated with various health benefits.

Table 3.5 (Continued)

Food/sources	Compounds	Notes
Soybeans	Genistein	Genistein is an isoflavone compound found in soybeans that has HDAC inhibitory activity. Soybeans are also a reliable source of protein and have been associated with various health benefits.
Fatty fish	Omega-3 fatty acids	Omega-3 fatty acids found in fatty fish like salmon and mackerel have been shown to have HDAC inhibitory activity. Fatty fish are also a good source of protein and have been associated with various health benefits.

3.6 Dietary Modulators of ncRNAs

The use of dietary agents that can promote anticarcinogenic ncRNA expression or suppress their pro-oncogenic ability. ncRNAs regulate nearly all biological processes; by silencing oncogenes and upregulating tumour suppressor gene expression, both miRNA and lncRNA can contribute to cancer initiation, promotion and progression (Table 3.6).

This table provides a summary of some of the known dietary modulators of non-coding RNAs involved in epigenetic regulation and cancer prevention. It is not an exhaustive list, and the effects of these dietary components may depend on dose, duration and timing of exposure, as well as individual factors such as genetics and other lifestyle factors.

3.7 Precautions and Issues with Dietary Chemoprevention

DNA hypomethylation is flexible. A diet low in sources of methyl groups results in DNA hypomethylation due to impaired SAM synthesis. SAM is the methyl donor for methyltransferases. Dietary sources of methyl donors include folate, methionine, betaine, choline and Vit B12. GSH Glutathione (GSH) synthesis can impair SAM synthesis and perturb DNA methylation. GSH is used to conjugate chemicals and is part of the physiological response to excrete foreign bodies and offers chemical protection. Upon exposure to environmental chemicals, GSH synthesis is enhanced (GSH). So, with the increasing toxin, there is

Table 3.6 ncRNA, dietary modulators and association with cancer.

Non-coding RNA	Dietary modulators	Mechanism of action	Cancer types
MicroRNA-29	Green tea polyphenols	Downregulate *DNMT3A* and *3B*	Breast, liver, colon
MicroRNA-34a	Resveratrol	Inhibit HDACs	Breast, colon, lung, prostate
MicroRNA-103	Omega-3 fatty acids	Upregulate tumour suppressor genes	Colon
MicroRNA-126	Sulforaphane	Inhibit *DNMT1*	Colon, prostate
MicroRNA-148a	Genistein	Inhibit *DNMT1*, *HDAC1* and *2*	Breast, prostate
MicroRNA-152	Curcumin	Inhibit *DNMT1*, *HDAC1* and *2*	Breast, colon, lung, prostate
MicroRNA-200c	Green tea polyphenols	Inhibit *EZH2* and *DNMT1*	Breast, lung
Long non-coding RNA H19	Curcumin	Downregulate *DNMT1*	Breast
Long non-coding RNA HOTAIR	Epigallocatechin-3-gallate (EGCG)	Inhibit *PRC2* and *EZH2*	Breast, liver, prostate
Long non-coding RNA MALAT1	Quercetin	Inhibit *EZH2*	Breast, liver
Long non-coding RNA MEG3	Sulforaphane	Downregulate *DNMT1*	Breast, colon, liver, prostate
Long non-coding RNA XIST	Genistein	Downregulate *DNMT1* and *HDAC2*	Breast

increasing GSH. If GSH-depleted homocysteine is shunted to GSH instead of methionine, resulting in decreased methionine, this then results in decreased SAM and consequentially decreased DNA methylation. Tolerance to transient exposure to toxins is due to homeostatic buffering. However, chronic exposure leads to depletion of intracellular GSH (through conjugation with toxin). Depleted GSH thus then results in DNA hypomethylation. Therefore, both dietary insufficiency of methyl groups and chronic low-level toxin exposure can result in methylation imbalance. It is logical, then, that there is increased hypomethylation seen with aging. GSH decreases due to a greater cumulative burden of chemicals and increased persistence of organic pollutants (synergistic effect). Chemical-induced changes are heritable; extreme nutrition-induced changes are heritable.

3.8 Epigenetics and Inflammation/Immune Response

It is believed that gene regulation of pathways associated with inflammation is at least in part epigenetically controlled by global DNA hypomethylation associated with inflammation. As epigenetics is strongly associated with molecular homeostasis and immune cells are crucial in repair and restoration, there is a logical connection between the two. There remains much work to be done in this field, especially expanding to histone modification in inflammation markers and validation in different tissues. Increasing understanding of epigenomic control of inflammatory markers is important clinically with regards to potential pharmaceutical targets for chronic conditions as well as potential biomarkers to be used in the prediction of inflammation-related disorders. The key roles of the immune system are not only in the identification of non-self but also in regulating tissue repair and resolution following injury. Injury can in this instance be an umbrella term encompassing trauma, toxins, emotional stress, mechanical damage, dramatic dietary changes and temperature fluctuations. Post-injury, immune cells react via signalling to the immune system to induce the appropriate injury adaptation response. The immune system then regulates, monitors and repairs to resolve the threat and restore homeostasis. If unsuccessful, inflammation becomes ongoing and chronic, potentially leading to disease rather than restoration of homeostasis.

Genome-wide changes in epigenetic mechanisms are associated with immune cell development throughout haematopoiesis and activation. Once acquired such changes persist throughout cell divisions. However, the reversibility of epigenetic marks means that there is plasticity in the transcriptional program of a given cell, even if it is already differentiated. Epigenetic mechanisms therefore are the ideal tool for controlling immune cell activity, which is highly specialised but also highly responsive to environmental stimuli (Table 3.7).

3.9 Epigenetic Inheritance Mechanisms

Sexual dimorphism refers to the differences between males and females of the same species, not only in terms of their physical characteristics but also in their behaviour, physiology and other traits. The mono-allelic expression of certain genes dependent on the parent of origin is referred to as genomic imprinting. In this case, a specific methylation pattern occurs during gametogenesis at specific genomic loci and then is maintained throughout adulthood. During embryogenesis, the correct expression from each parental contribution is required. Sexual dimorphism has been observed in DNA methylation patterns. For example, some genes may be differentially methylated between males and females due to differences in hormonal

Table 3.7 Epigenetic mechanisms targeted in treating immune disorders.

Epigenetics and inflammation/ immune disorders	Description	Examples
DNA methylation	Methylation of CpG sites in gene promoters can lead to decreased gene expression, and changes in methylation status have been linked to immune-related diseases	In lupus, hypomethylation of the promoter region of the FCRL3 gene, which is involved in B-cell activation, has been observed.
Histone modifications	Changes in histone modifications can impact gene expression and have been linked to immune-related diseases	In rheumatoid arthritis, an increase in histone deacetylase activity has been observed, which leads to a decrease in acetylation of histones and subsequent suppression of gene expression.
Non-coding RNAs	miRNAs and lncRNAs have been implicated in regulating immune cell differentiation and function, and changes in their expression have been linked to immune-related diseases	In psoriasis, miR-31 has been shown to regulate T-cell differentiation and modulate cytokine production. In asthma, lncRNA GAS5 has been shown to regulate T-cell apoptosis.
Epigenetic therapies	Targeting epigenetic enzymes or pathways has emerged as a promising approach for treating immune-related diseases	For example, the HDAC inhibitor vorinostat has been shown to suppress inflammation in rheumatoid arthritis and lupus. The DNMT inhibitor decitabine has been investigated for the treatment of lupus and other autoimmune diseases.

and physiological processes. This can impact various traits, including sexual development, behaviour and susceptibility to certain diseases.

Sexual dimorphism can influence the patterns of histone modifications, leading to differences in gene regulation between males and females. For instance, some genes involved in sex determination and hormone signalling pathways may exhibit distinct histone modification patterns. Non-coding RNAs, like microRNAs and long non-coding RNAs, can regulate gene expression post-transcriptionally. These molecules can exhibit sexual dimorphism in their expression levels and functions, leading to differences in gene regulation between the sexes. Non-coding RNAs can impact various biological processes, including development, metabolism and disease susceptibility.

Additionally, environmental factors, such as nutrition, stress and exposure to toxins, can influence epigenetic marks and can also contribute to sexual dimorphism. These factors can impact the epigenetic regulation of genes involved in various physiological and developmental processes, leading to differences between males and females. It is important to note that the interplay between genetics, epigenetics and environmental factors is complex and can vary between species and individuals. Epigenetic marks provide a potential avenue for understanding how sex-specific traits and behaviours are regulated at the molecular level and how they interact with genetic and environmental factors.

3.10 X-inactivation

In mammals, females have two X chromosomes, while males have one X and one Y chromosome. To balance gene dosage between the sexes, one of the X chromosomes in females undergoes X-chromosome inactivation (XCI). This process involves the silencing of most genes on one of the X chromosomes. XCI is a classic example of epigenetic regulation that leads to sexual dimorphism in gene expression. XCI is a fundamental epigenetic process that occurs in females to ensure the equal expression of X-linked genes between males and females, despite females having two X chromosomes and males having one X and one Y chromosome. XCI involves the silencing of most genes on one of the X chromosomes in each female cell, creating a dosage compensation mechanism. In early development, both X chromosomes in female cells are initially active. However, around the time of implantation in the uterus, one of the X chromosomes is randomly chosen to become inactive in each cell. This choice is made by the expression of the X-inactive specific transcript (*XIST*) gene located on the X chromosome to be inactivated. *XIST* produces a long non-coding RNA molecule that coats the chosen X chromosome, initiating a cascade of epigenetic modifications that lead to its silencing. The process of XCI involves a series of epigenetic modifications that alter the chromatin structure of the inactive X chromosome. DNA methylation patterns change on the inactive X chromosome, leading to increased methylation of CpG islands. This contributes to gene silencing by preventing the binding of transcription factors and other factors necessary for gene expression. The inactive X chromosome undergoes changes in histone modifications, including increased levels of repressive marks like H3K27me3 and decreased levels of active marks like H3K4me3. These modifications influence the three-dimensional structure of chromatin and further contribute to gene silencing. The inactive X chromosome becomes more compacted, forming a structure known as a Barr body. This compacted structure physically separates the inactive X from the active X and contributes to its transcriptional silencing.

While most genes on the inactive X chromosome are silenced, some genes manage to escape XCI and remain active. These genes are often found in regions called 'escapee' or 'pseudo autosomal' regions. The reasons for the escape of certain genes from XCI are not fully understood but are thought to involve specific chromatin configurations and epigenetic modifications that prevent their complete silencing. XCI has important implications for development and disease. In some cases, XCI can be skewed, where a higher proportion of cells in a female silence the same X chromosome. This can lead to conditions where certain mutations on the active X chromosome result in more severe disease symptoms than if they were on the inactive X chromosome. Additionally, disruptions in the XCI process can contribute to various genetic disorders and syndromes. For example, defects in the *XIST* gene or other factors involved in XCI can lead to disorders where both X chromosomes remain active, resulting in developmental abnormalities. In summary, XCI is a complex and highly regulated epigenetic process that ensures proper gene dosage between males and females. It highlights how epigenetic mechanisms, including DNA methylation, histone modifications and chromatin structure, play a critical role in regulating gene expression during development and across different cell types.

3.11 Genomic Imprinting

As outlined above, genomic imprinting is an epigenetic phenomenon that leads to the differential expression of genes based on their parental origin. In other words, certain genes are expressed or silenced depending on whether they are inherited from the mother or the father. This process is crucial for normal development and plays a significant role in various aspects of human biology. Imprinting disorders arise when there are disruptions in the normal epigenetic marks associated with imprinted genes. These disorders can lead to various health problems and developmental abnormalities. The evolutionary purpose of genomic imprinting is still debated. One theory is the 'parental conflict' hypothesis, where imprinted genes are involved in a competition between paternal and maternal alleles to optimise the allocation of resources to offspring. This hypothesis suggests that different alleles are favoured depending on their parent of origin, leading to the observed patterns of gene expression. Imprinted genes are critical for normal development and have important implications for human health when their epigenetic regulation is disrupted. In humans, there are over 200 known imprinted genes scattered across the genome. These genes are involved in a wide range of biological processes, including embryonic development, placental function, growth regulation and metabolism. Some well-studied imprinted genes in humans include: *IGF2, H19, UBE3A and SNRPN*.

IGF2 is a gene that encodes a protein involved in cell growth, differentiation and development. It plays a critical role in foetal and placental growth. In a normal scenario, the *IGF2* gene is typically more highly expressed by the paternal allele, while the maternal allele is silenced or expressed at a lower level. This is in contrast to *H19*, which is located adjacent to *IGF2*. The *H19* gene is typically expressed by the maternal allele, while the paternal allele is silenced. The regulation of *IGF2* and *H19* expression involves a region of DNA called the IGF2/H19 imprinting control region (ICR). This region contains DNA methylation marks that differ between the maternal and paternal alleles. The methylation pattern at the ICR is responsible for establishing and maintaining the imprinted expression patterns of *IGF2* and *H19*. The mechanism underlying this imprinting involves a complex interplay of DNA methylation, chromatin structure and regulatory elements. The IGF2/H19 ICR acts as a regulatory switch that controls the expression of both *IGF2* and *H19* in an allele-specific manner. Imprinting of IGF2 is essential for normal foetal growth and development. Disruptions in the normal imprinting of *IGF2* can lead to various developmental disorders, such as Beckwith–Wiedemann syndrome (BWS), where there is an overgrowth of foetal tissues and an increased risk of certain tumours.

H19 encodes a non-coding RNA molecule. It is typically expressed from the maternal allele, while the paternal allele is silenced or expressed at a lower level. The regulation of *H19* expression involves the same genomic region, the IGF2/H19 ICR, that is responsible for regulating *IGF2* expression. The imprinted expression of *H19* is crucial for normal embryonic development, including proper growth and differentiation. Disruptions in the imprinting of *H19*, like disruptions in *IGF2* imprinting, can lead to developmental disorders such as BWS and Silver–Russell syndrome.

The *UBE3A* (Ubiquitin Protein Ligase E3A) gene is another example of a gene that undergoes genomic imprinting in human embryonic development. *UBE3A* is located on chromosome 15 and is associated with the neurodevelopmental disorder known as Angelman syndrome (AS) when its imprinting is disrupted. In the case of *UBE3A*, the gene is primarily expressed from the maternal allele, while the paternal allele is silenced or expressed at a very low level in certain regions of the brain. This gene expression pattern is regulated by an imprinting centre located within the 15q11-q13 chromosomal region, where *UBE3A* is situated. The mechanism of *UBE3A* imprinting involves a few different steps, in neurons, the paternal allele of *UBE3A* is silenced, and an antisense transcript (UBE3A-ATS) is transcribed from the paternal allele. This transcript, also known as UBE3A-ATS or UBE3A antisense transcript, overlaps with UBE3A and helps to maintain the paternal allele's silencing. The antisense RNA produced by UBE3A-ATS can target and cause the degradation of UBE3A-specific RNA. This contributes to keeping the paternal allele's *UBE3A* expression low. The maternal allele, on the other hand, escapes this antisense-mediated silencing and is expressed in neurons. This differential expression is thought to be important for normal neural development and function. Disruptions in *UBE3A* imprinting are associated with AS. AS is a complex neurodevelopmental disorder characterised by severe cognitive deficits, speech impairment and motor dysfunction. Most cases of AS are caused by the loss of the maternal allele's expression due to deletions, mutations, or epigenetic changes in the *UBE3A* gene or its imprinting centre.

SNRPN (Small Nuclear Ribonucleoprotein Polypeptide N) is located on chromosome 15 within the same region as the *UBE3A* gene, specifically within the 15q11-q13 chromosomal region. *SNRPN* is primarily expressed from the paternal allele, and its expression is regulated by DNA methylation marks at an imprinting centre in the 15q11-q13 region. This region contains differentially methylated regions (DMRs) that play a crucial role in establishing and maintaining the imprinted expression pattern of *SNRPN*. In normal development, the paternal allele of SNRPN is expressed in a variety of tissues. This expression is driven by the epigenetic marks at the imprinting centre on the paternal chromosome. The maternal allele of *SNRPN* is typically silenced due to DNA methylation at the imprinting centre. This methylation prevents the expression of *SNRPN* from the maternal allele. In some cases of Prader–Willi syndrome, the paternal allele's expression is lost due to a genetic or epigenetic abnormality in the imprinting centre or surrounding regions. This leads to the absence of *SNRPN* expression from the paternal allele, contributing to the features of PWS. The imprinting centre in the 15q11-q13 region contains DMRs that play a pivotal role in controlling the allele-specific expression of *SNRPN* and other genes in the region. These epigenetic marks are established during gametogenesis and are important for maintaining proper gene expression patterns in embryonic development. Disruptions in the imprinting of *SNRPN*, either through loss of paternal expression or other alterations in epigenetic marks, can contribute to Prader–Willi syndrome, which is characterised by developmental and behavioural abnormalities as well as hypothalamic dysfunction leading to hyperphagia and obesity (Table 3.8).

3.12 Transgenerational Epigenetic Inheritance

The idea that experienced-induced phenotypes can be transmitted across generations originated with Lamarck in the early nineteenth century and his theories of use/disuse and the inheritance of acquired characteristics, which at the time were widely disregarded in favour of Darwin's theory of evolution and natural selection. The role of epigenetic processes in evolution is still questionable, but what is true is that the plasticity/homeostatic flexibility that epigenetic marks confer have especially important implications for furthering our understanding of the

Table 3.8 Some known human imprinted genes along with their associated imprinting disorders.*

Imprinted gene	Chromosome	Imprinted allele	Expression	Associated imprinting disorder
IGF2	11	Paternal	High	Beckwith–Wiedemann Syndrome (BWS)
H19	11	Maternal	High	Beckwith–Wiedemann Syndrome (BWS)
UBE3A	15	Maternal	High	Angelman Syndrome (AS)
SNRPN	15	Paternal	High	Prader–Willi Syndrome (PWS)
KCNQ1OT1	11	Maternal	High	Beckwith–Wiedemann Syndrome (BWS)
CDKN1C	11	Paternal	Low	Beckwith–Wiedemann Syndrome (BWS)
GNAS	20	Maternal	Low/ Paternal	Pseudohypoparathyroidism Type 1A (PHP1A)
PEG10	7	Paternal	High	None (No known imprinting disorder)
MEG3	14	Maternal	High	None (No known imprinting disorder)
NNAT	20	Paternal	High	None (No known imprinting disorder)

* Please note that this table is not exhaustive and includes only a selection of well-known imprinted genes. Additionally, the expression levels listed as 'High' or 'Low' indicate the relative expression from the imprinted allele. Some genes are associated with imprinting disorders, while others are not known to be associated with specific imprinting-related conditions. The field of genomic imprinting is complex and continues to be an area of active research, so the list of known imprinted genes may expand over time.

developmental origins of health and disease, and whilst not responsible for large-scale evolutionary events, they very plausibly contribute to homeostatic fluctuations of populations over time, responding to different circumstances. Enrichment of positive environmental factors has shown that learning and memory impairments can be overcome and then persist over subsequent generations in experiments with mice (Arai et al. 2009). Therefore, increasing our understanding of inheritance to include both genetic and epigenetic effects may provide better insight into effective therapeutic approaches and societal consequences, bringing together biomedical science and social science to improve health.

Transgenerational epigenetic inheritance refers to the transmission of epigenetic information from one generation to the next. This process involves the transfer of changes in gene expression patterns that are not caused by alterations in the DNA sequence itself but rather by modifications to the epigenetic marks, such as DNA methylation and histone modifications. While the mechanisms and extent of transgenerational epigenetic inheritance are still being explored, evidence suggests that it can occur in both humans and various other organisms. Research on transgenerational epigenetic inheritance in humans is still in its preliminary stages, and many aspects are not fully understood. However, there is evidence suggesting that certain environmental factors experienced by one generation can influence the epigenetic marks of their offspring. For example, studies have suggested that maternal nutrition during pregnancy can impact the epigenetic marks of the developing foetus. Poor maternal nutrition has been associated with changes in DNA methylation patterns in genes related to growth, metabolism and development. Prenatal stress experienced by pregnant women has been linked to alterations in DNA methylation patterns in genes involved in stress response and neural development. These changes may potentially affect the offspring's susceptibility to stress-related disorders. Exposure to environmental toxins, such as certain chemicals and pollutants, has been associated with transgenerational changes in DNA methylation and gene expression patterns. These changes may influence the risk of various diseases in offspring.

Key features of epigenetic mechanisms are their plasticity but also the potential stability of the marks, meaning that there is potential for epigenetic changes to be both persistent and heritable. Experiences across the human lifespan have the potential to induce changes to the epigenome; these changes have important implications for the neurobiology, physiology and behaviour of an organism, contributing to divergent developmental outcomes. There is increasing data as to the parental influence on the epigenome, resulting in changed gene expression and behaviour. Maternal effects are more studied, but there is increasing evidence of the paternal contribution to developmental regulation in the offspring. Together, this provides insight into the role of epigenetics as the mediating mechanism through which environmental experiences are transmitted between generations. There are considerable implications of environmentally induced transgenerational effects, particularly in how we assess risk, health and equity. These themes will be discussed further as we progress more into the details of the epigenetic impacts of stress, environment, behaviour and lifestyle (Table 3.9).

Epigenetic marks appear to maintain environmentally induced changes to phenotype both within and across generations, meaning a single genotype has the potential for multiple phenotypic outcomes, which attributes an organism with a high level of developmental plasticity. With regards to experience-dependent epigenetic modifications, such as nutritional programming of foetal metabolism, it may be the case that there are adaptive consequences of early life experience.

Table 3.9 Studies on transgenerational epigenetic inheritance.

Type of study	Organism	Exposure and generation(s) studied	Key findings
Famine study	Humans	Dutch Hunger Winter (1944–45) exposed pregnant women and their offspring	Lower DNA methylation of imprinted genes in offspring, associated with an increased risk of metabolic disorders
Maternal care study	Rats	High-licking/grooming versus low-licking/grooming mothers	Increased methylation of glucocorticoid receptor promoter in offspring of low licking/grooming mothers, associated with decreased gene expression and increased stress responses
Environmental toxin study	Rats	Bisphenol A (BPA) exposure during pregnancy in F0 generation	Altered DNA methylation in sperm of F1 generation, associated with behavioural and metabolic changes in subsequent generations
Behavioural study	Mice	Fear conditioning of F0 generation	Increased histone acetylation of memory-related genes in sperm of F1 generation, associated with improved fear memory retention
Maternal stress study	Mice	Maternal restraint stress during pregnancy	Transgenerational transmission of altered DNA methylation and histone modifications, associated with changes in stress responses and behaviour in subsequent generations
In vitro study	Human cells	In vitro treatment with histone deacetylase (HDAC) inhibitors	Altered gene expression and histone acetylation patterns in subsequent generations of cells, suggesting potential for transgenerational epigenetic inheritance

The thrifty phenotype arises when the prenatal period is characterised by under-nutrition and then 'reprograms' the individual to be conservative with energy output to promote the storage of glucose, this adaptation has adverse health consequences if there is a mismatch between food availability/quality in the prenatal and postnatal period.

Similarly, the experience could be related to stress and heightened HPA reactivity, typically elevated stress response is seen as a negative and associated with increased susceptibility to psychiatric and physical disease, but evolutionary speaking, heightened stress response could be advantageous in terms of being able to respond rapidly to a threat. Therefore, the consequences of early life experience can be considered as adaptive or maladaptive dependent on the match between early and later environmental conditions. Epigenetic marks have an integral role in enabling these homeostatic phenotypic adaptations.

Transgenerational epigenetic inheritance has been studied more extensively in other organisms, particularly in model organisms like mice, rats and plants. Studies have shown that exposure to dietary and environmental factors in rodents can lead to changes in DNA methylation patterns that are transmitted to subsequent generations. These changes can influence traits such as behaviour, metabolism and disease susceptibility. Plants have been shown to exhibit transgenerational epigenetic inheritance in response to various stressors, including temperature fluctuations and pathogen exposure. These inherited epigenetic changes can affect growth, development and stress tolerance in offspring. The roundworm Caenorhabditis elegans has provided insights into transgenerational epigenetic inheritance. For instance, exposure to starvation conditions in one generation can lead to changes in DNA methylation patterns that affect the lifespan of subsequent generations.

The mechanisms underlying transgenerational epigenetic inheritance are not fully understood, but several proposed mechanisms include sperm and egg epigenomes whereby epigenetic information can be stored in the germline cells (sperm and eggs) and passed on to the next generation. This can involve modifications to DNA methylation, histone modifications and small RNA molecules. Epigenetic marks can influence the three-dimensional structure of chromatin, affecting gene accessibility. Changes in chromatin states can persist through cell divisions and influence gene expression patterns in subsequent generations. Small RNAs, such as microRNAs, can also play a role in transmitting epigenetic information between generations. These molecules can affect gene expression by targeting specific mRNA molecules for degradation.

It is important to note that not all epigenetic changes are stably inherited across multiple generations, and the extent of transgenerational inheritance can vary depending on factors like the timing and type of exposure. In summary, transgenerational epigenetic inheritance is a complex and emerging field of research that suggests that environmental influences experienced by one generation can potentially affect the epigenetic landscape and gene expression patterns of subsequent generations. While evidence for transgenerational epigenetic inheritance exists in humans and other organisms, more research is needed to fully understand the mechanisms and implications of this process.

3.13 Conclusion

It is clear we need to re-evaluate what we understand about inheritance. There is increasing evidence that supports epigenetic modifications that maintain environmentally induced phenotypic variations within and across generations. A single genotype therefore has multiple phenotypic outcomes determined by a range of factors prior to, during and post-conception. This conveys considerable developmental and through-life plasticity, unlike random genetic mutation, which is associated as a consequence of adaptation to circumstances as opposed to random non-direct effect. The thrifty phenotype and nutritional programming are good examples of an adaptive response to early life experience. With undernutrition associated with programming to conserve energy output and promote glucose storage when there is a mismatch between prenatal and postnatal nutritional environments, this pre-programming can have negative health consequences. Similarly, heightened HPA reactivity – chronic stress and deprivation – potentially leads to an increased stressed response. The consequences of early life experience can be adaptive or maladaptive depending on how closely matched prenatal and postnatal environments. Broadening our understanding of inheritance to include genetic and epigenetic mechanisms may help to really understand the human condition and design effective therapeutic approaches.

Task

Design an experiment that would investigate either pharmaceutical or environmental factors on epigenetic mechanisms.

- *Given what is known about your topic, how can exogenous factors contribute positively or negatively to your problem?*
- *Can these factors be controlled for in a study?*
- *What is the appropriate study design, can you model your research hypotheses in cell lines or animal models?*

References

Arai JA, Li S, Hartley DM, Feig LA. Transgenerational rescue of a genetic defect in long-term potentiation and memory formation by juvenile enrichment. *J Neurosci.* 2009 Feb 4;29(5):1496–502. doi: 10.1523/JNEUROSCI.5057-08.2009. PMID: 19193896; PMCID: PMC3408235.

The article investigates the effects of juvenile environmental enrichment on the transgenerational rescue of a genetic defect in long-term potentiation (LTP) and memory formation in mice. The study found that exposing juvenile mice to an enriched environment led to improvements in LTP and memory formation in both the juvenile mice and their offspring, even when the offspring carried a genetic defect associated with impaired LTP and memory. The authors discuss the potential implications of these findings for understanding the mechanisms underlying cognitive function and the potential for environmental interventions to improve cognitive outcomes in individuals with genetic predispositions to cognitive impairments.

He Y, de Witte LD, Houtepen LC, Nispeling DM, Xu Z, Yu Q, Yu Y, Hol EM, Kahn RS, Boks MP. DNA methylation changes related to nutritional deprivation: a genome-wide analysis of population and in vitro data. *Clin Epigenetics*. 2019 May 16;11(1):80. doi: 10.1186/s13148-019-0680-7. PMID: 31097004; PMCID: PMC6524251.

The article investigates the effects of nutritional deprivation on DNA methylation patterns using a genome-wide analysis of population and in vitro data. The study identified significant DNA methylation changes in response to nutrient deprivation, particularly in genes related to metabolic processes and immune function. The authors discuss the potential implications of these findings for understanding the molecular mechanisms underlying the effects of malnutrition on health and disease, as well as the potential for epigenetic changes to be used as biomarkers of nutritional status.

Further Reading

Arora I, Tollefsbol TO. Computational methods and next-generation sequencing approaches to analyze epigenetics data: profiling of methods and applications. *Methods*. 2021 Mar; 187:92–103. doi: 10.1016/j.ymeth.2020.09.008. Epub 2020 Sep 14. PMID: 32941995; PMCID: PMC7914156.

The review article provides an overview of various computational and NGS methods used to study epigenetics. The authors discuss the strengths and limitations of different techniques, including DNA methylation profiling, chromatin immunoprecipitation sequencing (ChIP-seq) and RNA sequencing. They also highlight the bioinformatics tools and pipelines available for data analysis, including quality control, alignment, peak calling and differential analysis. The article emphasises the importance of integrating multiple epigenetic data types and using appropriate statistical methods for accurate interpretation of results. The authors conclude by discussing some of the current challenges in the field and future directions for epigenetic research using computational and NGS approaches.

Bheda P, Schneider R. Epigenetics reloaded: the single-cell revolution. *Trends Cell Biol.* 2014 Nov;24(11):712–23. doi: 10.1016/j.tcb.2014.08.010. Epub 2014 Oct 3. PMID: 25283892.

This paper discusses the recent advancements in single-cell epigenetics research. They discuss various techniques for profiling epigenetic modifications at the single-cell level, including bisulphite sequencing, ChIP-seq and single-cell RNA-seq. The authors highlight the importance of studying epigenetic heterogeneity at the single-cell level, as this can reveal new insights into cellular differentiation, disease progression and therapeutic response. The review also discusses the challenges associated with single-cell epigenetic studies, such as technical variability, low input amounts and the need for specialised computational methods. Finally, the authors suggest potential future directions for single-cell epigenetic research, including the integration of multiple modalities of single-cell data and the development of new technologies for epigenetic editing at the single-cell level.

Holdgate GA, Bardelle C, Lanne A, Read J, O'Donovan DH, Smith JM, Selmi N, Sheppard R. Drug discovery for epigenetics targets. *Drug Discov Today.* 2022 Apr;27(4):1088–98. doi: 10.1016/j.drudis.2021.10.020. Epub 2021 Oct 30. PMID: 34728375.

This paper discusses the current state and prospects of drug discovery for epigenetic targets. It provides an overview of epigenetic mechanisms and their role in diseases, such as cancer and neurological disorders. The authors highlight the challenges in targeting epigenetic proteins and pathways, such as the lack of selectivity and efficacy of current drugs, and provide insights into novel strategies, including the use of small-molecule inhibitors and immunotherapies. The article emphasises the need for interdisciplinary collaborations, advanced screening technologies and a better understanding of epigenetic mechanisms to develop effective and safe epigenetic drugs for clinical use.

Meaburn E, Schulz R. Next generation sequencing in epigenetics: insights and challenges. *Semin Cell Dev Biol.* 2012 Apr;23(2):192–9. doi: 10.1016/j. semcdb.2011.10.010. Epub 2011 Oct 19. PMID: 22027613.

This paper provides an overview of next-generation sequencing (NGS) technologies and their applications in epigenetic research. They discuss the distinct types of epigenetic marks and the sequencing technologies used to analyse them, as well as the challenges associated with NGS data analysis, including data processing, alignment and quality control. The authors also describe the applications of NGS in epigenetic research, including the identification of DMRs and the characterization of histone modifications. They conclude by discussing future developments and challenges in the field, including the need for standardised protocols and the integration of several types of omics data to provide a more comprehensive understanding of epigenetic regulation.

Miranda Furtado CL, Dos Santos Luciano MC, Silva Santos RD, Furtado GP, Moraes MO, Pessoa C. Epidrugs: targeting epigenetic marks in cancer treatment. *Epigenetics.* 2019 Dec;14(12):1164–76. doi: 10.1080/15592294.2019.1640546. Epub 2019 Jul 13. PMID: 31282279; PMCID: PMC6791710.

This article discusses the role of epigenetic modifications in cancer and the potential of epigenetic drugs (epidrugs) as a new class of anticancer agents. The authors provide an overview of epigenetic mechanisms, including DNA methylation, histone modifications and non-coding RNA, and their involvement in cancer development and progression. They also discuss the different classes of epidrugs such as DNA methyltransferase inhibitors, HDAC inhibitors and RNA interference agents, and their mechanisms of action. The article highlights the preclinical and clinical studies of epidrugs in several types of cancer and their promising results. The authors also discuss the challenges and limitations of epidrug therapy, such as drug resistance and off-target effects, and suggest potential strategies to overcome these limitations. The article concludes by emphasising the potential of epidrugs as a novel approach to cancer treatment and the need for further research to fully understand their therapeutic potential.

Peedicayil J. Pharmacoepigenetics and pharmacoepigenomics: an overview. *Curr Drug Discov Technol.* 2019;16(4):392–9. doi: 10.2174/1570163815666180419154633. PMID: 29676232.

The article provides an overview of pharmacogenomics and pharmacogenetics, two emerging fields of study that focus on understanding how an individual's genetic makeup and epigenetic changes influence their response to drugs. The author discusses various epigenetic mechanisms that can alter gene expression and drug metabolism, including DNA methylation, histone modification and non-coding RNA. The article also highlights the potential applications of pharmacogenomics and pharmacogenetics in personalised medicine, drug discovery and drug safety. Finally, the author addresses some of the challenges and limitations of these fields, including the need for more research on epigenetic changes and their effects on drug responses, as well as ethical concerns related to genetic testing and personalised medicine.

4

Tissues and Methods for Epigenetic Analyses

EWAS (epigenome-wide association studies) are becoming increasingly used due to substantial reductions in cost and data processing time. In such studies, the DNA methylation profiles of clinical samples are analysed to increase understanding of underlying pathogenic mechanisms. The two big challenges with EWAS are, as mentioned before, the problem of epigenetic marks having tissue specificity – this means clinically relevant samples need to be taken whereby tissue DNA is extracted that is relevant to the clinical condition being investigated. The other big challenge relates to the chosen method to employ to carry out the analysis; the big deciding factor as to what to choose here is weighing the balance between genomic coverage and cost-effectiveness, as this can have a significant impact upon the findings of the study. For some conditions, there are ethical and practical limitations in terms of the sample collected for analysis; for example, CNS disorders, this has meant that increasingly buccal and blood samples are used. As blood profiles are dominated by leucocyte DNA and are more sensitive to external influences such as infection, buccal samples are often now preferred due to their non-invasive nature and homogeneity in terms of cellular composition.

This chapter will briefly address the relative merits and weaknesses of different methods employed to investigate a range of epigenetic marks. Methods developed to study methylation include using methylation-sensitive restriction enzymes, bisulphite conversion and affinity enrichment using antibodies that are specific to 5-methylcytosine. Each method has its limitation: restriction enzymes result in a small subset of methylation sites, as they only identify those sites that are recognised by the restriction enzyme. Antibody-based affinity enrichment is biassed towards sites that have relatively high levels of cytosine methylation, that is, largely CpG islands. These methods were coupled with microarray-based methods to carry out genome-wide studies of DNA methylation and histone modifications based on the ChIP-chip method. The limitation here is that they do not truly provide comprehensive analysis throughout the whole genome, as synthesising

Epigenetics and Health: A Practical Guide, First Edition. Michelle McCulley.
© 2024 John Wiley & Sons, Inc. Published 2024 by John Wiley & Sons, Inc.

probes requires a priori knowledge of the region to be targeted. In contrast, sequencing-based approaches such as Chromatin immunoprecipitation followed by sequencing (ChIP-Seq) should, in theory, capture all DNA fragments isolated by immunoprecipitation if the sequencing depth/coverage is sufficient. Current DNA methylation microarrays can only look at a fraction of the DNA methylome. High-density microarrays such as the Infinium Human methylation 450 BeadChip allow the investigation of >450 000 CpG sites out of the potential approximately 28 million CpG sites in the human genome (0.02%), so they still suffer the downside of being restricted to preselected CpG sites and hence reduce the power of discoveries. Methylation in the non-CG context, which is revealed when sequencing the entire DNA methylome, would be overlooked by microarrays.

The rapid development and evolution of high-throughput sequencing technologies means that it is possible to sequence an exceptionally large number of sequence reads in parallel. This means that much larger-scale DNA methylation studies can be performed. For example, Illumina sequencing platforms are therefore advantageous for ChIP-Seq experiments that need a high coverage of sequencing reads to detect the enriched DNA fragments specific to a particular DNA modification after immunoprecipitation. Using the Illumina platform, results have suggested coverage of 94% of all cytosines in the genome through the generation of approx. 90 GB of sequencing data (Lister and Ecker 2009) (Table 4.1).

Table 4.1 Table summarising common methods used to analyse the epigenome.

Method	Description	Advantages	Disadvantages
Bisulphite sequencing	Sequencing-based method that determines DNA methylation status	High accuracy and sensitivity, can detect methylation at single-base resolution	Requires a large amount of input DNA, can be costly
ChIP-seq	Sequencing-based method that identifies genomic regions associated with specific histone modifications or transcription factors	High resolution and specificity, can identify binding sites and associated genes	Can require a large amount of input material, may not always distinguish between direct and indirect binding
ATAC-seq	Sequencing-based method that identifies open chromatin regions, which can indicate active regulatory elements	Requires a relatively small amount of input material, can detect active regions with high sensitivity	May not be able to distinguish between distinct types of regulatory elements

Table 4.1 (Continued)

Method	Description	Advantages	Disadvantages
DNA methylation microarrays	Array-based method that measures DNA methylation levels at specific loci	High throughput and relatively low cost, can analyse many samples in parallel	Limited to predefined loci, may not be able to detect methylation changes at single-base resolution
RNA-seq	Sequencing-based method that measures gene expression levels, which can be influenced by epigenetic modifications	High sensitivity and accuracy, can detect differential gene expression	Can be influenced by other factors besides epigenetic modifications, such as RNA stability
Methyl-CpG binding domain (MBD) sequencing	Sequencing-based method that enriches methylated DNA fragments using an MBD protein	High specificity and sensitivity, can identify differentially methylated regions	Limited to methylated DNA, can be influenced by nonspecific binding
Immunoprecipitation-based methods	Methods that use antibodies to enrich for histone modifications or other chromatin-associated proteins	Can provide high specificity and sensitivity for specific modifications or proteins	May not be able to distinguish between direct and indirect binding, can be influenced by antibody specificity

4.1 Methods for Assessing Genome-Wide DNA Methylation-EPIGENOMICS

There are many methods that can analyse differential methylation at the highest resolution of the single-base nucleotide. Broadly, these can be grouped as microarray and NGS sequencing-based approaches. Microarray-based technologies include the Infinium HumanMethylation27, Infinium 450K and Infinium methylationEPIC. These approaches use a fixed number of probes to survey specific genomic loci across the genome. The advantages of using Infinium 450K mean it has been the most widely used approach to study EWAS; these are low cost, modest DNA requirement and reduced sample prep time, all of which are advantageous to high-throughput analyses of large clinical numbers. The disadvantages to these approaches are that you are confined by the limitations of the probes used; the number and specificity of the probes will determine the findings, and information outside of these cannot be obtained with this approach.

NGS are important tools in epigenomics research for their ability to sequence a vast number of sequencing reads in parallel. Of the NGS approaches, WGBS (whole genome bisulphite sequencing) is widely regarded as the 'gold standard' because it provides the greatest genomic coverage. The significant cost and associated lengthy processing time mean that this approach is not practical for EWAS studies involving large numbers of samples. RRBS is an NGS-based approach developed as an alternative to WGBS to reduce time and cost; this method uses cytosine methylation-specific restriction enzyme digestion to focus specifically on CpG-rich regions. MeDIP-Seq is an affinity-enrichment platform that allows broader coverage of the genome than RRBS and uses methylation-specific antibodies to extract methylated cytosine-containing DNA fragments after DNA fragmentation. The disadvantage of MeDIP compared with RRBS is that it does not allow resolution at the base-pair resolution. Both RRBS and MeDIP are biassed towards CpG-rich repeats and are limited to specific regions because of the availability of the restriction enzymes or antibodies used. Another affinity enrichment platform that could potentially be a cost-effective method to use for EWAS is Methylation Capture Sequencing (MC-seq). This approach avoids the limitations of lower genome coverage found with Infinium 450K, a cheaper and quicker alternative to the costly WGBS, avoids the overrepresentation of CpG repeats associated with RRBS and methylated regions with MeDIP. MC-Seq uses target-specific bait sequences, utilising a targeted next-generation sequencing approach, but still needs to be explored further for reliability and reproducibility in clinical samples and in sensitivity to inter-individual variation.

4.2 Methods for Assessing Genome-Wide Histone Modifications

Up until the arrival of NGS, the majority of genome-wide epigenetic studies were dependent on DNA microarray methods such as ChIP-chip (chromatin immuno-precipitation based on microarray hybridisation of immune-precipitated DNA fragments). Sequencing-based methods such as ChIP-Seq have largely replaced microarray experiments for studying histone modifications and sequencing of the DNA methylome after bisulphite conversion has replaced microarrays targeting preselected CpG sites. Chromatin immunoprecipitation (ChIP) is a well-established technique used to study histone modifications and chromatin structures. The method utilises antibodies that recognise specific histone modification markers. For a specific target, ChIP followed by PCR (or qRT_PCR) can be used to identify specific histone modifications to the specific target DNA region. For undefined analyses, a sequencing approach is employed, such as ChIP-chip or ChIP-seq, to analyse histone modifications simultaneously at a vast number of loci.

Histone H3 is one of the DNA-binding proteins found in the chromatin of all eukaryotic cells. H3 along with four core histone proteins binds to DNA forming the structure of the nucleosome. The N-terminal tail of histone can undergo several distinct types of post-translational modifications that influence cellular processes. These modifications include the attachment of methyl or acetyl groups to lysine and arginine amino acids and the phosphorylation of serine or threonine.

Three histone H3-borne marks are known to be associated with gene expression. They respectively mark promoters (H3K4me3), distal regulatory elements (H3K4me1) and the active forms of both promoters and enhancers (H3K27ac). To quantify changes in these marks, a ChIP-based approach could be employed. ChIP followed by PCR allows relative quantification of the amount of protein at a targeted residue that is associated with a specific gene promoter region under various conditions. Briefly, differentiated cells are cross-linked, lysed and sonicated to generate DNA fragments with an average size of 200 bp. The cell lysate is then subjected to immunoprecipitation with antibodies against acetylated H3, methylated H3 or directly used as ChIP input. Enrichment of H3ac and H4ac determined by specific ChIP primers designed to amplify proximal sequences from the transcription start site of candidate genes. The analysis of bound DNA sequences through real-time PCR provides an accurate determination of levels of specific DNA in ChIP-ed samples.

Illumina TruSeq ChIP-Seq; ChIP DNA library prep and sequencing; this is high-resolution, genome-wide with no inherent bias associated with probe-based technology and captures histone modifications across the entire genome. As before, cells are cross-linked, lysed and sonicated to generate DNA fragments. The lysate is then immunoprecipitated with antibodies against acetylated H3 or methylated and used to construct sequencing libraries by ChIP-Seq Sample Prep Kit (Illumina). Enriched DNA sequencing is then performed on an Illumina Hiseq 2000 sequencer. Illumina and Life Technologies sequencing platforms can generate several hundred million short reads (<150 bp) providing them advantageous for ChIP-Seq where a high coverage of sequence reads is needed to detect the enrichment of DNA fragments specific to a particular histone modification after immunoprecipitation.

4.3 Integrative Analysis – Looking at DNA Methylation and Histone Modification

Trying to understand the dynamic interaction between methylation and chromatin is challenging but important in trying to elucidate the communication that occurs between DNA methylation and histone modifications or transcription factors that are methylation-sensitive in their role of regulation of gene transcription. ChIP-BMS has been developed for the detection of locus-specific methylation, and BisChIP-seq focuses on global profiling. ChIP-BMS determines methylation on ChIP DNA pulled down by a specific histone marker antibody, whereas BisChIP is based on sequencing of bisulphite-treated chromatin immunoprecipitated DNA.

4.4 Novel Technologies

Sequencing of DNA methylation using NGS still relies on bisulphite conversion to distinguish methylated from unmethylated cytosines. Third-generation sequencing technologies offer the potential advantage of directly sequencing the methylated cytosines without the need for bisulphite conversion. The issue with bisulphite conversion and NGS is that it is not possible to distinguish between 5mC and 5hmC; additionally, methylation outside of CG methylation is not studied, for example, 5-methyladeninine or methylation of other nucleotides. Both are likely to provide greater insight into human biology and disease. Novel methods to specifically detect 5hmC in genomic DNA include GLIB, which uses both enzymatic and chemical steps to isolate DNA containing 5hmC through the addition of glucose to each 5hmC. The other approach converts 5hmC to CMS by treating with sodium bisulphite and then immunoprecipitating CMS (cytosine 5-methylenesulphonate) containing DNA with a specific antiserum to CMS.

Sodium bisulphite conversion of DNA followed by sequencing is still the gold standard for detecting cytosine methylation. The shortcomings are the inability of this method to distinguish between 5mC and 5hmC; therefore this approach cannot detect methyladenine and thirdly, the bisulphite treatment causes damage to the DNA, which limits the method to sequence shorter DNA sequences and makes it difficult to study a haplotype or specific patterns of methylation in the parental chromosomes. Third-generation sequencing and single-molecule real-time sequencing overcome these problems as there is no need for bisulphite conversion. These methods rely on monitoring the kinetics of incorporated nucleotides in newly synthesised DNA strands by DNA polymerase. Nanopore sequencing can detect methylated cytosine directly. These methods also require much less starting material than NGS platforms.

4.5 Single-Cell Approaches

More recently, there has been interest in single-cell sequencing approaches for epigenomics to understand specific footprints of cellular differentiation. Single-cell DNA methylation can be analysed using single-cell bisulphite sequencing and single-cell reduced representation bisulphite sequencing. Single-cell ChIP-seq can be done using Drop-ChIP. CUT&Tag can also be employed to profile chromatin. In this method, an antibody is used to identify a target chromatin protein, such as a histone modification. Next, protein A and Tn5 transposase fusion proteins bind to the antibody and are tagged to the genomic regions where the target protein is bound. The single-cell platforms C1 and Chromium both enable single-cell ATAC-seq. Single-cell bisulphite sequencing provides insight at a single-cell level; however, as previously mentioned, there are DNA degradation issues due to the treatment with bisulphite, low input and restricted ability to assess heterogeneity in methylation between individual cells. This can be corrected using PBAT, which uses an adaptor tag allowing low-input DNA to be PCR amplified prior to deep sequencing. Both scWGBS and scRRBS are used for single-cell methylation analysis. This can be particularly useful to look at cell function during embryonic development and in tumour ontogeny. Ideally, this could be integrated with other single-cell technologies such as transcriptomics and metabolomics to explore the functional and dynamic links in a complex system.

PacBio single molecule real-time (SMRT) and nanopore are third-generation sequencing technologies that result in long sequencing reads and have no requirement for bisulphite conversion. Nanopore sequencing detects DNA methylation patterns via deviation in the ion current as each base is sequenced through the pore. This is best suited to bulk samples rather than tiny amounts of DNA (no PCR amplification). The major advantage of SMRT is that it can detect both nucleotide sequence and major types of methylation patterns (5mC, 5hmC, 6mA 4mC) simultaneously through monitoring the kinetics of the polymerase during the synthesis of dsDNA from fluorescently labelled nucleotides. Disadvantages of this method are that there can be inaccurate readings due to variable sensitivities to different methylation patterns; typically, 6mA and 4mC produce a kinetic signal that has high confidence, whereas the 5mC signal is low, and this makes this technology more suited to analysis of bacterial genomes.

4.6 ncRNAs

Important regulatory ncRNAs are miRNA, siRNA, piRNA and lncRNA, where lncRNA is normally more than 200 nt. Methods for ncRNA are well established and include high-throughput RNA sequencing, hybridisation-based microarrays and quantitative reverse transcription PCR, for which many kits are available. To look at targets of ncRNAs there are other specific techniques such as HITS-CLIP for identifying functional protein–RNA interaction sites and LIGR-seq for looking at miRNA-mRNA interactions.

For lncRNAs, there are many high-throughput techniques, that have been utilised based on the desired outcome. ChIRP and CHART for exploring lncRNA binding sites on chromatin genome-wide, RAP for mapping lncRNA, RIP, RIP-chip, RIP-seq for protein–RNA interaction, RNA pull-down and LC-MS/MS to identify proteomic interaction of a target lncRNA.

4.7   Oxidised DNA Methylation

Assessing oxidised forms of DNA methylation is useful if you want to assess DNA demethylation, the process mediated by TET enzymes that remove the methyl group from 5mC and make oxidised forms of 5mC, the most abundant being 5hmC, found to have a key regulatory role in many pathological and physiological process including tumorigenesis and embryogenesis. 5hmC can be detected using oxidative bisulphite sequencing (OxBS-seq), a modified bisulphite sequencing technique. Because this requires multiple bisulphite treatments causing DNA degradation, this technique will typically need a higher amount of input DNA and higher sequencing depths to be able to confidently determine low-frequency modifications. hmTOP-seq (Gibas et al. 2020) is a high-resolution bisulphite-free method recently developed to map whole genome 5hmC content.

4.8   Data Analysis

Key issues in the analysis of epigenomic data include large datasets, identification of general pipeline, data processing and quality control, alignment to reference genome, quantification and visualisation of DNA methylation, profiling, validation and interpretation. In terms of raw data, for array-based technologies e.g. Illumina450K the output is relative quantification of the fluorescence of methylated and unmethylated loci. In the case of bisulphite sequencing-based WGBS and RRBS, the methylation status is determined at the single nucleotide level by comparing differential methylation loci (DML) and differential methylated regions (DMR). There are many bioinformatic tools that can assist with this, e.g. Bismark is used for bisulphite sequencing read alignment, DMAP can be used to identify differentially methylated regions and a web-based genome browser such as UCSC can be used for data visualisation. For ChIP-seq analysis, the pipeline includes data processing, read mapping, peak calling and visualisation, quantitative assessment of changes and functional analysis. DROMPAplus is a universal ChIP-seq pipeline tool. Further developments are critical in handling increasingly large and complex datasets and the integration of multi-omic data.

4.9   Conclusion

There are a range of methods available to study epigenomics. The method chosen will depend not only on the research question but also on cost and accessibility to various techniques and expertise. When thinking about your own project, the following questions are useful to consider helping determine the appropriate tissue, tools and method of data analysis to use.

Task

Use the questions below to guide what approach would be best to answer your research question:

- *How can you examine whether there is an epigenetic contribution to your research issue? What are you wanting to look at?*
- *What organism are you studying?*
- *What is already known, and what methods have been used before?*
- *What is the starting material for your study? What is a relevant tissue?*
- *What is your budget, and what scale do you want to take in your approach targeted or whole genome?*
- *What can you do? Do you want to look at a gene or genome level? If so, what is possible, and what resources do you need?*
- *Is DNA methylation relevant or histone modification?*
- *5mC or 5hmC??*
- *Is single-cell analysis useful?*

The table below summarises some potential scenarios and methods best suited to each case.

Scenario	Recommended technique	Issues to be aware of	Alternative techniques
• Limited budget • I want to look at DNA methylation • I know the gene/targeted region of interest	PCR-based bisulphite sequencing	• Inconsistency due to DNA template degradation after bisulphite treatment • Need >10 replications to confirm results	• Methylation-specific PCR (MSP) • Pyrosequencing

(*Continued*)

Scenario	Recommended technique	Issues to be aware of	Alternative techniques
• Unlimited budget • I want to explore genome-wide methylation/no a priori target of interest • I would like single-base resolution • My sample is human	Whole genome bisulphite sequencing (WGBS)	• Expensive • Needs sophisticated bioinformatic analysis • Large amount of input DNA required	• Human 450 – only human samples, cost-effective, coverage dependent on a predesigned array • RRBS – only 1–3% of the genome, covers most representative CpG island in gene regulatory regions, can detect DNA methylation in different species, and may lose coverage at intergenic/distal regulatory regions, cost-effective • scWGBS or scRRBS – particularly useful for germ cells, embryonic cells, and cancer cells – individual information
• I want to explore genome-wide methylation • I want to avoid bisulphite treatment • I am interested in detecting different methylation patterns • My sample is bacterial	Single molecule real-time (SMRT) sequencing	Large input DNA required	• Nanopore; large input required, difficult to understand performance reliability across species and sequencing batches
• I am interested in exploring histone modification across the genome • I have a large funding budget and can access high-quality antibodies	Chromatin immuno-precipitation followed by sequencing (ChIP-seq)	• Expensive • Good-quality antibody required	• ChIP-chip – more cost-effective; microarray – targets certain genomic regions – custom-designed coverage, has typical issues associated with microarray e.g. probe performance/hybridisation efficiency • Lower budget

Scenario	Recommended technique	Issues to be aware of	Alternative techniques
			• ELISA-based assay – cost-effective, commercial kits available but only gives a rough estimation of total changes of specific histone modification – broad prediction of histone modification changes • ChIP-PCR – detects enrichment of specific histone modification – only detects abundance, not single nucleotide resolution

The following website by the International Human Epigenome Consortium provides tools for analysis, publicly accessible data sets and information on reference epigenome standards for epigenomic analysis. This is a good place to start thinking about a research project from the perspective of understanding what has already been done, what data sets are available and the tools to be used in the analysis. http://ihec-epigenomes.org/.

References

Gibas P, Narmontė M, Staševskij Z, Gordevičius J, Klimašauskas S, Kriukienė E. Precise genomic mapping of 5-hydroxymethylcytosine via covalent tether-directed sequencing. *PLoS Biol*. 2020 Apr 10;18(4):e3000684. doi: 10.1371/journal. pbio.3000684. PMID: 32275660; PMCID: PMC7176277.

The article presents a new method for the precise genomic mapping of 5-hydroxymethylcytosine (5hmC), an important epigenetic modification that has been implicated in various biological processes and diseases. The method involves the use of a covalent tether to direct the sequencing of 5hmC-containing DNA molecules, allowing for high-resolution mapping of 5hmC at single-base resolution. The authors demonstrate the utility of this method in characterising 5hmC patterns in various tissues and cell types and show that it can be used to distinguish between 5hmC and 5-methylcytosine, another common DNA modification. The authors suggest that this method will be valuable for understanding the role of 5hmC in gene regulation and disease, as well as for developing new epigenetic diagnostic and therapeutic tools.

Lister R, Ecker JR. Finding the fifth base: genome-wide sequencing of cytosine methylation. *Genome Res.* 2009 Jun;19(6):959–66. doi: 10.1101/gr.083451.108. Epub 2009 Mar 9. PMID: 19273618; PMCID: PMC3807530.

The article discusses the emergence of the field of epigenomics, specifically the use of genome-wide sequencing to identify and analyse cytosine methylation patterns, which are referred to as the 'fifth base' of DNA. The authors provide an overview of the different technologies used to detect cytosine methylation and the challenges associated with analysing epigenetic modifications at a genome-wide scale. They also discuss the potential implications of cytosine methylation for understanding gene expression, development and disease, as well as the potential for epigenetic modifications to be used as diagnostic and therapeutic targets.

Further Reading

Angarica VE, Del Sol A. Bioinformatics tools for genome-wide epigenetic research. *Adv Exp Med Biol.* 2017;978:489–512. doi: 10.1007/978-3-319-53889-1_25. PMID: 28523562.

This article reviews bioinformatics tools that are used to study epigenetic modifications on a genome-wide scale. The authors provide an overview of the different types of epigenetic modifications and the tools available for their detection and analysis. They discuss various methods for analysing ChIP-seq and DNA methylation data, including peak calling, differential analysis and pathway enrichment analysis. The article also covers the use of machine learning techniques to identify epigenetic biomarkers and predict gene expression from epigenetic data. Finally, the authors highlight the importance of integrating diverse types of epigenetic data to gain a comprehensive understanding of epigenetic regulation in various biological processes.

Gouil Q, Keniry A. Latest techniques to study DNA methylation. *Essays Biochem.* 2019 Dec 20;63(6):639–48. doi: 10.1042/EBC20190027. PMID: 31755932; PMCID: PMC6923321.

This article provides an overview of the latest techniques to study DNA methylation. The authors discuss the principles, advantages and limitations of different methods such as bisulphite sequencing, reduced representation bisulphite sequencing, single-cell methylome analysis and high-throughput sequencing-based assays. The review also highlights the importance of integrating different methods to obtain a comprehensive understanding of DNA methylation patterns and their functional relevance in health and disease.

5

Normal and Abnormal Epigenetic Variation

In this chapter, we will consider the broad question of epigenetic variation. Unlike genetic variation, epigenetic variation changes within an individual over time and in response to environmental stimuli. We will focus on complex conditions, inflammation, immunity and environmental exposure to toxins.

Researchers are increasingly coming to the realisation that exploring epigenetics in conjunction with complex human diseases is especially important. The success of GWAS in identifying thousands of genetic variants and loci for complex traits and diseases still only confers a modest effect size, typically with an odds ratio of less than 1.5, meaning these variants only account for a small contribution to the heritability of complex phenotypes. Epigenetic mechanisms are believed to be responsible for the missing remaining contribution to heritability (via germline).

As a major regulator of gene expression, the epigenome is responsive to a broad range of environmental factors, including toxin exposure, diet, stress and socioeconomic circumstances. Epigenomic changes occur throughout the lifespan and are sensitive to environmental influences including exercise, stress, diet and shift work. The epigenome therefore provides a homeostatic mechanism at the molecular level that allows phenotypic malleability in response to changing internal and external environments. Age-related methylation changes are part of a complex process contributing to normal variation; there are also ageing and disease-related changes, such as cancer, Parkinson's disease, Alzheimer's disease, autoimmunity and diabetes.

Age-related methylation and epigenetic drift are terms used to refer to the changes in DNA methylation that occur with ageing; both are affected by lifestyle and exposure to environmental factors. Longitudinal studies of identical twins have demonstrated age-related discordance in DNA methylation patterns. Some age-related methylation changes have a clear functional purpose, for example, the shutting down of developmental genes. On the other hand, age-related changes

Epigenetics and Health: A Practical Guide, First Edition. Michelle McCulley.
© 2024 John Wiley & Sons, Inc. Published 2024 by John Wiley & Sons, Inc.

are enriched in pathways that involve stem cell differentiation. Even though the processes involved in epigenetic reprogramming are tightly regulated, 5-mC levels have been shown to change during early development, adolescence and adulthood in response to environmental exposures. The environment is central to shaping the changing epigenome throughout human ageing. Environment and lifestyle drive age-specific changes in DNA methylation status, not simply genetics.

DNA methylation is the most studied epigenetic mark in ageing studies due to its stability and availability of high-throughput quantification methods. Consistent findings in studies of the ageing methylome are locus-specific increases in DNA methylation with age, global decreases in methylation with age and bidirectional changes in DNA methylation variability over time, i.e. hypomethylation or hypermethylation changes vary by specific gene regions. Additionally, DNA methylation specific to tissue type has also been found to vary with respect to ageing. Age-related methylation refers to predictable direction-specific changes in DNA methylation levels that happen with normal ageing. Age-related methylation is suggestive of meaningful changes that happen in homeostatic control as we age.

Age-related accumulation of errors in the epigenetic machinery associated with epigenetic drift leads to the gradual loss of methylation in hypermethylated regions and the gain of methylation in hypomethylated regions. Higher rates of drift are observed in proliferative tissues. DNMT1 often occupies hypermethylated gene bodies and bivalent chromatin; loss of DMNT1 binding can break down chromatin boundaries over time, resulting in the spread of methylation to gene promoters and loss of methylation in gene bodies. Additionally, sp1 and sp3 transcription factors and RNA polymerase II are associated with resistance to de novo methylation in promoter regions. DNA-protein interactions are also likely to influence age-related DNA methylation changes. Indirect mechanisms can also bring about the same effect; for example, decreased expression of DNA methyl transferases can be brought about as a consequence of heavy metal exposure. Inflammation and oxidative stress are also potential candidates for altering the epigenome through impacting the availability of methyl donors, inducing stem cell proliferation, etc. It is likely that there would be greater effects in time-critical periods of development whereby an individual is more susceptible to perturbations. Diet has been shown to impact the normal age-related methylation change trajectory. A 2014 study by Horvath et al. demonstrated that BMI is associated with accelerated age-related methylation at 353 CpG sites in the human liver, suggesting that nutritionally induced oxidative stress and metabolic alterations can alter the trajectory of normal age-related methylation changes. Stress has also been found to accelerate age-related DNA methylation changes in a 2015 study of soldiers returning from deployment in Afghanistan by Boks et al. (2015). Cumulative lifetime stress too has been associated with accelerated age-related DNA methylation changes.

Epigenetic drift, in contrast to age-related methylation, is less predictable and refers to bidirectional changes in epigenetic variability with age, which could have an impact on the overall methylome plasticity, a consequence of methylation maintenance failure in cellular replication. These are random inefficiencies that increase in likelihood because of ageing, such changes are unlikely to be consistent between individuals of a given population and thus are not predictive of age but more so a consequence of environmental influence on the epigenome, hence, explaining the discordance found between identical twins, which becomes increasingly divergent as the twins age.

MZ twins are epigenetically indistinguishable during the early years of life, whereas older MZ twins have major differences in their overall content and genomic distribution of 5mC and histone acetylation, indicating they inherit identical epigenetic profiles from their parents and then acquire different somatic epigenetic markers throughout their life suggesting there is a key role for epigenetics in the development of complex disease phenotypes.

It is an interesting question to pose whether epigenetic mechanisms play a role in the ageing process or are a consequence of human ageing. Given that ageing is a risk factor for many conditions, understanding the precise contribution of epigenetic mechanisms to ageing can help in understanding the pathogenesis of complex diseases and potentially provide novel biomarkers that can predict mortality risk. Molecular biomarkers investigated to predict risk include genomic instability, mitochondrial dysfunction, telomere attrition and cellular senescence. As discussed earlier, DNA methylation patterns change with age. The epigenetic clock is a method developed specifically using epigenetic biomarkers of age and refers to the DNA methylation age. This is used to provide an estimate of age from different tissues and different life stages with relative accuracy. An application of the epigenetic clock is to identify individuals that deviate significantly from normal ageing, that is their epigenetic clock is atypical for their chronological age. Research to date has identified individuals with an accelerated epigenetic clock and found positive associations with a range of age-associated diseases (including cancer, cardiometabolic disease, dementia and diabetes), as well as unhealthy behaviours and mortality risk. Validated epigenetic clocks used in such research include the Horvath clock, based on DNA methylation at 353 CpGs, and Hannum's clock based on DNA methylation at 71 CpGs. Research using the Hannum epigenetic clock suggests that there is an association between accelerated DNA methylation age and increased risk of mortality. Key limitations to be aware of when focusing on epigenetic clocks is the fact that DNA methylation levels are dynamic and are significantly affected by environmental factors such as stress and smoking (Table 5.1).

For studies of ageing, cohort studies would be preferred over case-control. Longitudinal studies would be the gold standard, following individuals over time

Table 5.1 Overview of life-course factors that influence the epigenome.

Life course event	Description	Example epigenetic changes
In utero development	Epigenetic changes that occur during foetal development and can affect long-term health outcomes	Altered DNA methylation of imprinted genes and histone modifications in pluripotent stem cells
Early life experiences	Epigenetic changes that occur during infancy and childhood and can affect later health outcomes	Altered DNA methylation of stress-related genes and histone modifications in brain development
Puberty	Hormonal changes associated with sexual maturation can affect the epigenome	Altered DNA methylation and histone modifications in reproductive tissues; changes in histone modifications in the brain
Ageing	Epigenetic changes that occur over the course of a lifetime and can affect health outcomes in old age	Global hypomethylation, promoter hypermethylation in tumour suppressor genes; changes in histone modifications in immune system regulation
Environmental exposures	Exposure to chemicals, pollutants and other environmental factors can affect the epigenome	DNA methylation changes in response to air pollution and histone modifications in response to endocrine disruptors
Diet and lifestyle	Diet and lifestyle factors, such as nutrition and physical activity, can affect the epigenome	DNA methylation changes in response to dietary methyl donors and histone modifications in response to exercise

and tracking their association with disease progression and having biological samples taken at multiple time points to elucidate whether DNA methylation drives ageing or vice versa.

Environmental exposure, immune response and lifestyle undoubtedly impact human disease risk. Here, we will focus on the impact these factors have at a molecular level through epigenetic marks.

5.1 Epigenetics and Trained Immunity

Trained immunity refers to the recent discovery of the memory properties of the innate immune cells (monocytes/macrophages and natural killer cells), whereby these cells can recollect a previous foreign encounter, priming their immunological response. This enhanced response is accompanied by changes in intracellular metabolism and epigenetic regulation at the level of histone modifications. The innate immune cells may also play a role in immune-mediated diseases, including inflammatory conditions and autoimmune disorders. Understanding the role of epigenetics in trained immunity can help improve vaccination strategies as well as help to understand and treat other immune-mediated diseases.

A key difference between acquired immunity and trained immunity is that whereas acquired immunity primes the immune response for an enhanced response to subsequent encounters with the same pathogen, the trained innate response instructs a future nonspecific response. Historically, the first evidence of trained immunity came from early data of BCG vaccination, whereby broad protection was conferred against unrelated diseases. Innate immune memory can arise from a diverse range of immunological challenges, and monocyte memories can result in vastly different responses to future exposures. Trained immunity induced by BCG, B-glucan or oxLDL defines an enhanced non-specific response to future infections by enhancing inflammatory and antimicrobial properties of innate immune cells. Conversely, stimulation with LPS induces a persistent refractory state known as tolerance with a markedly reduced capacity to respond to re-stimulation. Cellular memory is facilitated by major shifts in metabolic and transcriptional pathways that are bidirectionally linked through epigenetic mechanisms, supporting a prominent role for the epigenome in recording adaptive experiences. Stable and heritable epigenetic changes can explain long-term memory effects of trained immunity observed months after BCG vaccination.

Molecular mechanisms enable monocytes to incorporate micro-environmental experiences into their program of gene expression, a chromatin-based memory that, importantly, has the potential for reversibility of innate immune phenotypes. H3K4m3 has been associated with trained immunity. Several studies report on the enrichment of H3K4m3 promoters in genes encoding proinflammatory cytokines and intracellular signalling molecules after stimulation with B-glucan; the heterologous benefit of the BCG vaccine is associated with persistent enrichment at the promoters of genes encoding TNF-alpha, IL6 and TLR4. Less is known about the role of DNA methylation in innate immune memory. Epigenome-wide studies have shown a general loss of 5mc during ex vivo monocyte-to-macrocyte differentiation, most differentially methylated regions occurring at distal enhancers and only a small fraction at gene promoters.

5.2 Epigenetics and Vascular Senescence

As humans age, vascular ageing occurs via pathological processes that drive the changes accompanying vascular ageing, such as vascular cell senescence, inflammation, oxidation stress and calcification. As we age, these pathologies, driven by epigenetic alterations, gradually accumulate and are linked to various age-related diseases. Many abnormal DNA methylation patterns have been characterised in association with various age-related diseases. Enzymes that regulate histone modifications, including HDAC, are widely known to play a significant role in vascular ageing. ncRNAs are differentially expressed in senescent and normal cells. Typically, a person's inflammatory status increases with age, and DNA hypermethylation has been found to be correlated with chronic inflammation ageing-related diseases. Vascular ageing contributes to cardiovascular disease. Ageing is irreversible, but epigenetic alterations are reversible and may provide novel treatment targets for patients with age-related cardiovascular disease. Association between epigenetic regulation of multiple gene targets linked to vascular ageing, could be appropriate therapeutic targets for the management of vascular ageing and related diseases.

There is still a long way to go, and several key issues need to be addressed, for example, how to manipulate histone modifications in specific tissues, the problem being systemic inhibition or activation of HDAC and other epigenetic enzymes, which can result in adverse reactions because of their broad target range. Additionally, there is also a greater need to monitor epigenetic changes in cells, potentially using recent technologies of single-cell sequencing and single-cell epigenetic technologies such as scATAC-seq, scDNase-seq and scChic-seq.

5.3   Epigenetics and Obesity

Obesity is characterised by increased adipose tissue, and it is associated with an increased risk of chronic diseases such as T2DM, hypertension, cardiovascular disease, arthritis, infertility, mental health disorders and cancer. The energy imbalance resulting from increased intake of energy and decreased physical activity is a major contributing factor to obesity. However, it is important to note that obesity is complex and involves behavioural, genetic and environmental factors. The development of obesity is affected by many pathways that have a net result in metabolic dysfunction. Such pathways include hedonic and homeostatic eating behaviours regulated by the brain, tissue energy expenditure and changes in adiposity via differentiation of adipocytes and accumulation of lipids in adipose tissue. Many of these pathways are controlled by epigenetic mechanisms, and understanding precisely how they work has important implications for the treatment, prevention and diagnosis of obesity-related conditions.

Hypothalamic proopiomelanocortin (POMC) neurons control homeostatic eating by expressing appetite-inhibitory POMC. An obesogenic diet causes repression of *POMC* and results in increased food intake. In part, this is countered by acute POMC neuron plasticity change via upregulation of *ST8SIA4*. Adipose tissue secretes leptin to signal to the hypothalamus that there is excess energy storage; it does this through upregulation of POMC. An obesogenic diet numbs the leptin response in the POMC neurons and represses leptin transcription in adipocytes. Compulsive eating is promoted through diet-induced dysregulation of tyrosine hydroxylase, dopamine transporter and cyclin-dependent kinase inhibitor 1C in the dopaminergic neurons in the central reward circuitry resulting in a boosted desire for palatable food.

Adipocyte differentiation and fat accumulation can be upset through the action of nutrients, specifically targeting CEBPA, PPARG and fatty acid synthase causing adiposity. Dietary elements can also influence energy expenditure by regulating ornithine decarboxylase and spermidine-spermine *N*-acetyltransferase in fat and possibly in the liver through NNMT, which acts by inhibiting the enrichment of H3K4 methylation at ornithine decarboxylase and spermidine-spermine *N*-acetyltransferase. In adipose tissue, energy expenditure is regulated by ODC and SSAT via H3K4me3 and NNMT. Leptin is released and acts on the hypothalamus in the brain through MECP2, MBD2, DNMT1/3A, promoter DNA methylation, and miR-200a, -200b, -132, -145 and -429. Adiposity is governed by the regulation of CEBPA (CCAAT/enhancer binding protein [C/EBP] alpha), PPARG (peroxisome proliferator-activated receptor gamma) and FASN (fatty acid synthase) via H3K9me3 (SETDB1– SET domain bifurcated), H3K20me (SETD8- Set domain-containing protein 8), promoter DNA methylation and miR-27a, 27b.

In the brain, the central reward circuitry genes are involved in hedonic eating: *tyrosine hydroxylase, dopamine transport solute carrier family 6* and *cyclin-dependent kinase inhibitor 1C* through promoter methylation and *DNMT3A*. Hypothalamus genes are involved in energy homeostasis; *POMC* and *ST8 alpha-N-acetyl-neuraminide alpha2,8 sialyltransferase* are regulated through histone acetylation (KAT8-lysine acetyltransferase), *MECP2* (methyl CpG binding protein 2), promoter DNA methylation and miR-103. Energy expenditure in the liver is regulated by the genes *ODC* (ornithine decarboxylase) and *SSAT* (spermindine speririmine *N*-acetyltransferase) via H3K4me3 and NNMT (nicotinamide *N*-methyltransferase). A lot of work has been done on exploring the role of noncoding RNA in the epigenetic mechanisms contributing to obesity. miRNA has been shown to be significantly associated with leptin signalling and adipogenesis, challenges remain in identifying specific targets of miRNA within obesity-related molecular pathways, partly because miRNA-target binding patterns are very variable and partly because they are also tissue-specific.

5.4 Epigenetics and Cumulative Toxin Exposure: Toxicoepigenomics

Environmental epigenetics involves the ability of environmental contaminants to change epigenetic controls that then change genome activity in terms of up/down-regulating gene expression; it is through epigenetics that we can explain the mechanism through which our environment intimately interacts with our biology. In terms of specific environmental contaminants, adult exposure to butyl-paraben has been shown to produce DNA methylation changes in sperm, thus having potential for multigenerational transmission. One of the most visual effects that illustrate the epigenomic flexibility in response to interference is exemplified by the research carried out on mouse models where maternal dietary exposure to Bisphenol A (BPA) resulted in changes in coat colour and obesity status of their pups. The dietary intake of BPA reduced DNA methylation, which switched on the normally suppressed agouti gene, resulting in yellow-furred fat pups instead of normal-sized brown-fur pups. Even more interesting, the normal gene function could be restored through maternal dietary supplementation with methyl donors such as folate, choline, betaine and vitamin B12. Other studies on rodents have also demonstrated that both pre- and perinatal exposure to BPA can impact higher body weight, increased cancer susceptibility, altered reproductive function and other chronic health effects. More studies in humans are also beginning to emerge with BPA concentration correlated with obesity in children, and now a widespread hypothesis is that life course exposure to environmental pollutants is linked to adult disease.

Once a toxin has entered a living organism, it interacts with the organisms' cellular components and disturbs them, as in the case of humans, leading to health impairment. Heavy metals, pesticides and fertilisers are examples of widespread environmental pollutants that have serious connotations for human health including cancer, their effect occurs through interference with the normal regulation of our genome, altering which genes are switched on and off. The mechanisms through which environmental contaminants interact with our genome are likely to be explained through epigenetic control, this variability in switching genes on and off, interfering with the body's natural mechanism to respond to environmental stimuli when trying to maintain homeostasis. Analysis of the epigenome in a subset of the human population with known chronic, long-term exposure to such chemical stressors can provide novel insight into underlying mechanisms affecting long-term health. Abnormalities of DNA methylation have been well established in the pathogenesis of cancer. Environmental chemical exposure, particularly to polychlorinated biphenyls (PCBs) and pesticides has been significantly associated with non-Hodgkin's Lymphoma (NHL). Previous studies have found that individuals who have experienced chronic, longer-term exposure to pesticides from crop spraying, etc. have a higher incidence of NHL. Identification of an epigenomic profile of chronic toxin exposure could be useful in the early diagnosis, management and treatment of these individuals.

Exposure to environmental factors during critical developmental periods in life changes disease susceptibility later in life by impacting developmental plasticity. Through-life exposures exert their effects through altering gene regulation via changes in the epigenome, which result in the subsequent change to health phenotype. Environmental exposure can alter the rate of age-related methylation/ epigenetic drift. Greater exposure to toxicants will have a greater impact on age-related methylation changes and cause a greater shift away from the normal baseline of typical age-related methylation change. Toxicant exposure can bring about long-term changes to gene control via epigenetic marks. To test the impact of toxicant exposure, longitudinal studies are required, whereby there is sufficient early life exposure data and then repeated epigenetic assessment to identify loci that are specifically affected.

Prenatal toxicant exposure alters the establishment of DNA methylation marks during development in animal models; these environmental-induced changes have subsequently been shown to be carried over to adulthood. This suggests the possibility of prenatal environmental exposure potentially modifying age-related DNA methylation changes. Important considerations for the contribution of epigenetics to the developmental origin of health and disease and intergenerational transmission of environmental stressors are the characterisation of the toxicological stressor, the window of exposure and the target tissue and function. In terms of identifying the toxicological stressors, this includes whether this is a single stressor or multiple stressors acting together, such as a high-fat diet and exposure to toxic chemicals or low SES and high air pollution exposure. It is critical to incorporate epidemiological and experimental data to attribute cause. For the window of exposure, the regulation of epigenetic marks is different with respect to developmental stage, so the time in a person's life history when the exposure occurred can be of particular significance. Some exposures may be long-lasting and consistent and may have an impact on different epigenomic targets at different developmental stages, thus making inference more complex. The time at which an epigenetic marker is assessed may not reflect the timing of exposure, and clinical symptoms could manifest at a later point.

In terms of target tissue, blood, semen and placenta are most widely studied for larger epidemiological epigenetic studies; however, this may not represent processes in the target tissue, e.g. lung epithelium. There is still a lack of data correlating epigenetic markers in blood with those in target tissues. These fluids themselves are also very heterogeneous in terms of cell composition, etc. We also need to consider the specificity of epigenetic changes; for example, do different stressors lead to the same epigenetic changes? Alternatively, a single stressor could lead to a broad range of epigenetic effects. At the molecular level, why does exposure to the stressor lead to the modification of the epigenetic pattern of only a certain number of genes? This could be a combination of the generic effects of the stressor at a specific time window. We also need to be able to make intelligent interpretations of complex data, for example, what is the biological likelihood between epigenetic changes and health outcomes? This is difficult, as we need to know function/effect of each gene individually and collectively in the genome. It is also important to consider the long-term effects of stressors such as smoking, air pollution and endocrine disruptors.

In a 2016 study by Joubert et al., investigating the impact of exposure to tobacco smoke, the Cord blood from almost 7000 new-borns across 13 cohorts and 6000 differentially methylated CpG sites were analysed using Illumina 450K data. The key issue investigated here is whether methylation patterns in early life induced by environmental exposure persist to older ages. This study demonstrated that, in blood, this was the case, with DNA changes persisting at least to adolescence. Many of the genes identified have been associated with dysmorphology, cancer and asthma development, but it is not yet known whether these methylation changes and associated changes in gene expression contribute to the disease pathway or are simply biomarkers of exposure related to maternal smoking status. There are some studies to date that have suggested that DNA methylation changes in the blood are in fact also consistent with changes in the target tissue, which is the lung epithelium. Other studies have demonstrated that smoking has a broad impact on the methylome and that the associated changes remain long after smoking cessation.

Air pollution has been associated with methylation changes of genes related to oxidative stress, immunity and inflammatory responses. A 2017 study using NO_2 as an indicator of traffic pollution investigated the impact on methylation levels of the genome of newborns ($n = 1508$) and found less significant hits than the smoking studies but did highlight anti-oxidative defence genes. Other studies have found potential links with immune cell function, so there is a need now for larger studies to strengthen these findings. Endocrine-disrupting compounds (EDC) such as BPA, phthalates, metals and dioxins are also of interest in terms of the epigenetic changes they produce. There are currently relatively few human studies linking EDC exposure to epigenetic marks. From studies looking at the placenta, current research appears to suggest that environmental exposure, for example, to heavy metals, can disrupt the placental genome or epigenome leading to altered disease susceptibility. Such could be abnormal birth or adverse outcomes later in life, supporting the DOHaD hypothesis. Numerous animal and human studies focusing on the analysis of sperm have shown males can pass on phenotypic traits gained through dietary changes, chemical exposure, stress and trauma throughout their lives to their offspring. Transmission of epimutations from father to offspring must be mediated by sperm. Specific changes in sperm DNA methylation patterns that have retained histone profiles and small non-coding RNA transcriptomes have been identified in animal models of epigenetic transgenerational inheritance.

5.5 Epigenetics, COVID-19 and Environmental Chemical Exposures

COVID-19 was identified as a global pandemic and had widespread detrimental effects globally. The lethality of COVID-19 has been linked to cytokine storm, age and the immune system of an individual. Epigenetics of both the virus and the host can impact the severity of disease and infectivity of the virus. The SARS-CoV-2 virus enters the host cell through binding of the viral spike protein to the host receptor (ACE2) facilitating entry of the virus into the host cell via the use of other host proteins including furin and proteases. Once inside the host cells the virus multiplies using the host cellular machinery and then assembled and packaged for secretion as virions through exocytosis. Viruses can influence the host immune system and consequentially epigenetic changes that then either promote or suppress viral replication and spread.

Epigenetic susceptibility loci for respiratory failure in COVID-19 have been identified. These were specific DNA methylation signatures that were associated with patients needing hospitalisation for oxygen therapy. This resulted in the identification of an EPICOVID signature, this signature is linked to levels of pro-inflammatory cytokine IL-6, C-reactive protein, ferritin, fibrinogen and D-dimer and total lymphocyte count. The EPICOVID signature appeared to show specificity for COVID-19 infection and associated hyperinflammation and immune activation. There is potential for this kind of information to be used in a hospital laboratory setting to monitor patients and identify higher-risk patients. So, this could be an additional tool to fight COVID-19 in terms of allowing for better patient follow-up and restricting immune activation and organ failure seen in late disease. Further studies have taken this further and carried out transcriptome analyses to identify changes in gene transcription attributable to epigenetic regulation.

From the outset of the COVID-19 pandemic, there has been a rapid acceleration of research into potential risk factors for COVID-19. At present it is accepted that ageing, obesity, chronic inflammation, hypertension and other clinical factors contribute to increased risk of severity, in addition to these increasing data supports a role for environmental chemicals to impact susceptibility to infection and severity of symptoms. Large genome-wide studies have identified certain loci with increased susceptibility to severity of presentation with COVID-19. Epigenome-wide studies have also revealed differences in terms of methylation level differences at specific CpG sites associated with increased risk of hospitalisation and requirement for respiratory support, suggesting that there is an epigenetic contribution to the immunological response to SARS-CoV-2 infection. The mode in which environmental contaminants impact COVD-19 susceptibility is likely to be through epigenetic mechanisms, in particular the epigenetic regulation of the immune pathways employed in response to SARS-CoV-2.

Air pollution, toxic metals and metalloids, polyfluorinated substances and endocrine-disrupting chemicals are all capable of inducing epigenetic changes in the form of changes to DNA methylation, miRNAs, or histone modifications. Such epigenetic changes can impact then on immune pathways which in turn increase COVID-19 susceptibility or severity. Likely immune pathways affected through epigenetic alterations include viral entry, viral recognition, cytokine production and immunologic memory. There is still extremely limited data linking environmental chemicals to SARS-CoV-2 susceptibility and the severity/progression of COVID-19 but given that the impacts on human health of exposures are well known and documented, now more research is needed to understand the epigenetic mechanisms and immune pathways that contribute to increased susceptibility to adverse COVID-19 outcomes. This is particularly important to understand due to the potential reversibility of epigenetic mechanisms. To address this fully a greater need for diverse collaboration between different research disciplines.

5.6 Case Focus: Rheumatology

We can use rheumatology as an example of a complex disease whereby epigenetics contributes to the pathophysiology and where a greater understanding could lead to better treatment outcomes. As discussed previously, the epigenome is altered by endogenous and environmental factors and changes with age. Typically, although generalist, gene activation is associated with lower levels of methylation in gene promoter regions and distinct histone marks such as acetylation of amino acids in histones. A decrease in methylation is associated with ageing. Epigenetic mechanisms have been implicated in the pathogenesis of common rheumatic diseases including rheumatoid arthritis, osteoarthritis, systemic lupus erythematous and scleroderma. To date, EWAS studies in genetically complex inflammatory rheumatic diseases have identified correlations between epigenetic mechanisms and disease activity and severity. Epigenetic drift is associated with age-related changes, potentially increasing the risk for conditions such as polymyalgia rheumatic. Therapeutic targeting of the epigenome has shown promise in animal models of rheumatic disease. As discussed above, epigenetic mechanisms are essential for immune cell differentiation and function, including the correct activation of B cells and T cells and inflammatory processes. Systematic epigenomic screening can help classify and identify novel biomarkers for the personalised management of patients with inflammatory rheumatic diseases. The real potential of epigenetic markers is that these epigenetic alterations may be reversed or modified, whereas the genetic background is fixed; therefore, epigenetic markers are good therapeutic targets. HDAC inhibitors have an anti-inflammatory effect in various cell types, including RASF, and in animal models of RA. Other strategies include targeting downregulation of DNMT1 (and hypomethylation) in CD4 + cells, a potential target for SLE. This acts through a similar mechanism as methotrexate in that it reverses downregulation by increasing DNA hypermethylation. Therefore, in terms of designing treatments targeting the epigenetics of RA, pharmacological treatments can directly modify enzymes that catalyse epigenetic changes, target factors that indirectly affect global epigenetic profiles, and as such are HDAC inhibitors (HDAC increased in several inflammatory diseases), or promote upregulation of DNMT expression.

In terms of how epigenetic-focused research could aid in RA understanding, you could, for example, use baseline methylation as a marker of treatment response, i.e. methotrexate, or explore apparent inefficacy of anti-TNF agents. The differences in TNF promoter methylation status could be a potential reason no association has been found between TNF promoter polymorphism and response to anti-TNF antibody therapy. Additionally, disease-specific management, would include isoform selectivity of HDAC to make it more effective and safer (there are 11 isotypes HDAC1-HDAC11) and to directly/indirectly target the Ras-ERK-DNMT1 pathway.

The key application would be the rational use of epigenomic information in the clinical setting and in personalised medicine. This would entail the identification of epipolymorphisms associated with clinical outcomes, DNA methylation as a contributor to disease susceptibility in rheumatic conditions, the discovery of novel epigenetic mechanisms that modulate disease susceptibility and the development of new epigenetic therapies. As with other epigenetics studies, there are issues and limitations including the relevance of sample type/tissue, that complex techniques are needed for their determination, the potential link between epigenetic alterations and disease susceptibility SNPs in RA and the intelligent interpretation of genomic/epigenomic data. The major challenge is interpreting clinical relevance to enable prediction of the evolution of the disease, establish new treatments and address the development of personalised therapies.

5.7 Conclusion

The flexibility and dynamic responsiveness of the methylome demonstrate the important potential of lifestyle changes in rebooting epigenetic control of genes. This also makes the methylome an attractive biomarker to evaluate the molecular impact of lifestyle changes as well as a potential target for therapeutic intervention. Comparison of comprehensive methylation patterns in healthy individuals at multiple time points can further our understanding of whether such changes can be attributed to health and act as early indicators of chronic disease, present at a time point where there is significant potential to reverse effects. This also allows for the opportunity to establish what changes are normal and attributed to lifespan and what changes are related to disease progression. Key areas for future study are to focus on how age-related epigenetic changes contribute to genome maintenance in ageing with the aim of trying to prevent age-related diseases and potentially directly promote healthy longevity. It is highly likely that epigenetics provides the link between environmental exposure-related programming and intergenerational effects. More research is needed to understand the actual contribution of epigenetic regulation and associated mechanisms as well as the public health implications.

In conclusion, epigenomic characterisation can give rise to relevant epigenetic biomarkers indicative of disease type, prognosis and response to treatment. Unlike genetic mutations, epigenetic modifications are inherently reversible, making them attractive therapeutic targets. The challenge for translational research now is to understand how epigenetic alterations affect different cell types involved in pathogenesis using next-generation high-throughput methods of analysis and discover new ways to translate information into the clinical setting.

Task

Thinking about your research topic try to think about what the impacts of normal epigenetic variation could be on your study:

- *For the phenotype/condition you are interested in, is age likely to impact on the phenotype?*
- *If so, is this due to metabolic changes, changes in environmental exposures, hormones or another factor?*
- *If you model your conditon in an animal or cell-line model, how can you account for any age-related impacts?*
- *Does your study need a control population? How accurate would a control be? What would you need to account for? Is it appropriate for your study to have a control?*
- *Do you need to sample at multiple time intervals to allow for normal epigenetic variation?*
- *Is it possible to determine a baseline epigenome for your condition focused on a specific set of biomarkers which are more sensitive to epigenetic control?*

References

Boks MP, van Mierlo HC, Rutten BP, Radstake TR, De Witte L, Geuze E, Horvath S, Schalkwyk LC, Vinkers CH, Broen JC, Vermetten E. Longitudinal changes of telomere length and epigenetic age related to traumatic stress and post-traumatic stress disorder. *Psychoneuroendocrinology* 2015 Jan;51:506–12. doi: 10.1016/j. psyneuen.2014.07.011. Epub 2014 Jul 23. PMID: 25129579.

The article investigates the longitudinal changes in telomere length and epigenetic age in individuals exposed to traumatic stress and post-traumatic stress disorder (PTSD). The study found that individuals with PTSD had shorter telomere length and accelerated epigenetic ageing compared to controls, and that exposure to traumatic stress was associated with greater age-related changes in telomere length and epigenetic age. The authors suggest that these findings may have implications for understanding the biological mechanisms underlying the effects of traumatic stress on health and disease, and that telomere length and epigenetic age may serve as potential biomarkers of stress exposure and PTSD risk.

Horvath S, Erhart W, Brosch M, Ammerpohl O, von Schönfels W, Ahrens M, Heits N, Bell JT, Tsai PC, Spector TD, Deloukas P, Siebert R, Sipos B, Becker T, Röcken C, Schafmayer C, Hampe J. Obesity accelerates epigenetic aging of human liver. *Proc Natl Acad Sci U S A*. 2014 Oct 28;111(43):15538–43. doi: 10.1073/pnas.1412759111. Epub 2014 Oct 13. PMID: 25313081; PMCID: PMC4217403.

In this research, the authors investigated the impact of obesity on the aging of the human liver at the epigenetic level. They found that obesity appears to accelerate the process of epigenetic aging in the liver, as suggested by changes in DNA methylation patterns. The study provides valuable insights into the relationship between obesity and the aging of the liver and highlights the potential importance of epigenetic modifications in this process

Joubert BR, Felix JF, Yousefi P, Bakulski KM, Just AC, Breton C, Reese SE, Markunas CA, Richmond RC, Xu CJ, Küpers LK, Oh SS, Hoyo C, Gruzieva O, Söderhäll C, Salas LA, Baïz N, Zhang H, Lepeule J, Ruiz C, Ligthart S, Wang T, Taylor JA, Duijts L, Sharp GC, Jankipersadsing SA, Nilsen RM, Vaez A, Fallin MD, Hu D, Litonjua AA, Fuemmeler BF, Huen K, Kere J, Kull I, Munthe-Kaas MC, Gehring U, Bustamante M, Saurel-Coubizolles MJ, Quraishi BM, Ren J, Tost J, Gonzalez JR, Peters MJ, Håberg SE, Xu Z, van Meurs JB, Gaunt TR, Kerkhof M, Corpeleijn E, Feinberg AP, Eng C, Baccarelli AA, Benjamin Neelon SE, Bradman A, Merid SK, Bergström A, Herceg Z, Hernandez-Vargas H, Brunekreef B, Pinart M, Heude B, Ewart S, Yao J, Lemonnier N, Franco OH, Wu MC, Hofman A, McArdle W, Van der Vlies P, Falahi F, Gillman MW, Barcellos LF, Kumar A, Wickman M, Guerra S, Charles MA, Holloway J, Auffray C, Tiemeier HW, Smith GD, Postma D, Hivert

MF, Eskenazi B, Vrijheid M, Arshad H, Antó JM, Dehghan A, Karmaus W, Annesi-Maesano I, Sunyer J, Ghantous A, Pershagen G, Holland N, Murphy SK, DeMeo DL, Burchard EG, Ladd-Acosta C, Snieder H, Nystad W, Koppelman GH, Relton CL, Jaddoe VW, Wilcox A, Melén E, London SJ. DNA methylation in newborns and maternal smoking in pregnancy: genome-wide consortium meta-analysis. *Am J Hum Genet*. 2016 Apr 7;98(4):680–96. doi: 10.1016/j. ajhg.2016.02.019. Epub 2016 Mar 31. PMID: 27040690; PMCID: PMC4833289.

This article reports a large consortium meta-analysis that investigates the association between maternal smoking during pregnancy and DNA methylation in newborns. The study involved the analysis of genome-wide DNA methylation data from 13 227 individuals. The results showed significant associations between maternal smoking and DNA methylation changes in 6073 CpG sites, with differential methylation in genes related to developmental, regulatory and metabolic processes. The study provides important insights into the epigenetic consequences of maternal smoking during pregnancy and its potential effects on offspring health.

Further Reading

Ballestar E, Li T. New insights into the epigenetics of inflammatory rheumatic diseases. *Nat Rev Rheumatol*. 2017 Oct;13(10):593–605. doi: 10.1038/ nrrheum.2017.147. Epub 2017 Sep 14. PMID: 28905855.

This review highlights the role of epigenetics in inflammatory rheumatic diseases (IRDs). They describe how epigenetic alterations, such as changes in DNA methylation, histone modifications and non-coding RNAs, are involved in the pathogenesis of IRDs. The authors also discuss the potential use of epigenetic biomarkers for the diagnosis, prognosis and personalised treatment of IRDs. In addition, they describe how environmental factors, such as diet, smoking and pollution, can influence epigenetic modifications and contribute to the development of IRDs. Finally, the authors highlight the need for further research to fully understand the complex interplay between genetics, epigenetics and environmental factors in IRDs.

D'Aquila P, Rose G, Bellizzi D, Passarino G. Epigenetics and aging. *Maturitas*. 2013 Feb;74(2):130–6. doi: 10.1016/j.maturitas.2012.11.005. Epub 2012 Dec 12. PMID: 23245587.

The article provides an overview of the role of epigenetics in ageing, discussing how changes in epigenetic marks, contribute to age-related changes in gene expression and cellular function. The authors describe how environmental factors, such as diet, exercise and stress, can affect epigenetic modifications and influence the ageing process. They also highlight the potential of epigenetic interventions, such as dietary supplements and drugs, for promoting healthy ageing and preventing age-related diseases. The authors emphasise the importance of continued research in epigenetics and ageing to improve our understanding of the ageing process and develop effective interventions for healthy ageing.

Gladish N, Merrill SM, Kobor MS. Childhood trauma and epigenetics: state of the science and future. *Curr Environ Health Rep.* 2022 Dec;9(4):661–72. doi: 10.1007/ s40572-022-00381-5. Epub 2022 Oct 15. PMID: 36242743.

The article provides an overview of the current state of research on the relationship between childhood trauma and epigenetic modifications and discusses future directions for this area of study. The authors describe the various types of childhood trauma that have been linked to epigenetic changes, including abuse, neglect and poverty, and provide examples of the specific epigenetic modifications that have been observed in association with these experiences. The article also covers the mechanisms by which childhood trauma may lead to epigenetic changes, such as alterations in stress response systems and immune function. The authors discuss the implications of these findings for understanding the long-term impact of childhood trauma on mental and physical health and highlight the potential for epigenetic modifications to serve as biomarkers for trauma exposure. Finally, the article concludes by outlining future research directions, including the need for longitudinal studies to better understand the timing and persistence of epigenetic changes following childhood trauma, and the potential for interventions to reverse or mitigate these changes. Overall, the article provides a comprehensive overview of the current state of knowledge on the link between childhood trauma and epigenetics.

Gref A, Merid SK, Gruzieva O, Ballereau S, Becker A, Bellander T, Bergström A, Bossé Y, Bottai M, Chan-Yeung M, Fuertes E, Ierodiakonou D, Jiang R, Joly S, Jones M, Kobor MS, Korek M, Kozyrskyj AL, Kumar A, Lemonnier N, MacIntyre E, Ménard C, Nickle D, Obeidat M, Pellet J, Standl M, Sääf A, Söderhäll C, Tiesler CMT, van den Berge M, Vonk JM, Vora H, Xu CJ, Antó JM, Auffray C, Brauer M, Bousquet J, Brunekreef B, Gauderman WJ, Heinrich J, Kere J, Koppelman GH, Postma D, Carlsten C, Pershagen G, Melén E. Genome-wide interaction analysis of air pollution exposure and childhood asthma with functional follow-up. *Am J Respir Crit Care Med.* 2017 May 15;195(10):1373–83. doi: 10.1164/ rccm.201605-1026OC. PMID: 27901618; PMCID: PMC5443897.

The study investigated the interaction between air pollution exposure and genetic factors in childhood asthma. The researchers performed a genome-wide interaction analysis and found several significant interactions between air pollution exposure and genetic variants associated with asthma-related phenotypes. Functional follow-up experiments provided further evidence for the biological plausibility of the identified interactions. The study highlights the importance of considering gene-environment interactions in the development of childhood asthma and suggests potential targets for personalised prevention strategies.

Jeffries MA. The development of epigenetics in the study of disease pathogenesis. *Adv Exp Med Biol.* 2020; 1253:57–94. doi: 10.1007/978-981-15-3449-2_2. PMID: 32445091.

The article provides an overview of the role of epigenetics in disease pathogenesis, discussing how epigenetic mechanisms contribute to the development and progression of various diseases, including cancer, autoimmune disorders and neurological conditions. The author highlights the importance of understanding the complex interplay between genetics, epigenetics and the environment in disease pathogenesis. The article also covers the potential of epigenetic therapies for disease treatment and prevention, as well as the challenges and limitations in targeting epigenetic mechanisms. The author emphasises the need for continued research in epigenetics to improve our understanding of disease biology and develop personalised and effective treatments. Overall, the article is a valuable resource for researchers and clinicians interested in the role of epigenetics in disease.

King SE, Skinner MK. Epigenetic transgenerational inheritance of obesity susceptibility. *Trends Endocrinol Metab.* 2020 Jul;31(7):478–94. doi: 10.1016/j.tem.2020.02.009. Epub 2020 Mar 24. PMID: 32521235; PMCID: PMC8260009.

This article discusses the concept of epigenetic transgenerational inheritance, where epigenetic marks are passed down from one generation to the next, resulting in phenotypic changes in offspring without changes in DNA sequence. They focus on the inheritance of obesity susceptibility and discuss how environmental factors such as diet and exposure to endocrine disruptors can alter epigenetic marks and lead to transgenerational effects. The authors also discuss potential mechanisms for these effects, including changes in DNA methylation, histone modifications and non-coding RNA expression, and highlight the importance of understanding these mechanisms for future prevention and treatment strategies.

Klein K, Gay S. Epigenetics in rheumatoid arthritis. *Curr Opin Rheumatol.* 2015 Jan;27(1):76–82. doi: 10.1097/BOR.0000000000000128. PMID: 25415526.

This paper discusses the role of epigenetics in rheumatoid arthritis (RA), a chronic inflammatory autoimmune disease that affects the joints. The authors describe the distinct types of epigenetic modifications that have been observed in RA, including changes in DNA methylation, histone modifications and non-coding RNAs. They discuss how these modifications can influence gene expression and contribute to the development and progression of RA. The article also highlights the potential of epigenetic therapies as a novel approach to treating RA, and how they might be used in combination with traditional treatments. Overall, the article emphasises the growing importance of epigenetics in understanding the molecular mechanisms underlying RA and the potential for epigenetic interventions to improve patient outcomes.

de Lima Camillo LP, Quinlan RBA. A ride through the epigenetic landscape: aging reversal by reprogramming. *Geroscience*. 2021 Apr;43(2):463–85. doi: 10.1007/s11357-021-00358-6. Epub 2021 Apr 6. PMID: 33825176; PMCID: PMC8110674.

This article discusses the role of epigenetics in ageing and the possibility of reversing ageing through reprogramming. It covers various aspects of ageing and how they are influenced by epigenetic changes. The article discusses the use of reprogramming techniques, such as induced pluripotent stem cells (iPSCs), to reverse ageing and rejuvenate cells. It also explores the challenges associated with these techniques and the potential ethical considerations that may arise. Overall, the article provides a comprehensive overview of the epigenetic landscape of ageing and the emerging field of reprogramming as a potential tool to reverse ageing.

Oblak L, van der Zaag J, Higgins-Chen AT, Levine ME, Boks MP. A systematic review of biological, social, and environmental factors associated with epigenetic clock acceleration. *Ageing Res Rev*. 2021 Aug; 69:101348. doi: 10.1016/j.arr.2021.101348. Epub 2021 Apr 28. PMID: 33930583.

This article provides a systematic review of the biological, social and environmental factors associated with epigenetic clock acceleration. The authors discuss the concept of epigenetic ageing, or the gradual changes in DNA methylation patterns that occur with ageing, and review the methods used to measure epigenetic age, including epigenetic clocks. The article also reviews a range of factors that have been associated with accelerated epigenetic ageing, including genetic factors, lifestyle factors such as smoking and diet, and psychosocial factors such as stress and trauma. The authors highlight the need for further research to fully understand the complex interplay between these factors and their impact on epigenetic ageing and age-related diseases. The article concludes by discussing the potential implications of epigenetic ageing research for personalised medicine and disease prevention.

Orozco-Solis R, Aguilar-Arnal L. Circadian regulation of immunity through epigenetic mechanisms. *Front Cell Infect Microbiol*. 2020 Mar 13; 10:96. doi: 10.3389/fcimb.2020.00096. PMID: 32232012; PMCID: PMC7082642.

The article discusses the connection between the circadian rhythm and immune system regulation through epigenetic mechanisms. It highlights how the disruption of the circadian rhythm may lead to dysregulation of the immune system and the development of various diseases. The authors also explore the molecular mechanisms of how the circadian clock regulates the epigenetic modifications in the immune cells and how they, in turn, affect gene expression and immune responses. The article also discusses the potential therapeutic approaches targeting circadian rhythm and epigenetic regulation for the prevention and treatment of immune-related disorders.

Perera BPU, Faulk C, Svoboda LK, Goodrich JM, Dolinoy DC. The role of environmental exposures and the epigenome in health and disease. *Environ Mol Mutagen.* 2020 Jan;61(1):176–92. doi: 10.1002/em.22311. Epub 2019 Jun 20. PMID: 31177562; PMCID: PMC7252203.

This review article discusses the impact of environmental exposures on the epigenome and the resulting effects on human health and disease. The authors provide an overview of the epigenetic mechanisms, including DNA methylation and histone modifications and their role in regulating gene expression. They also discuss various environmental exposures that can lead to epigenetic changes, such as exposure to pollutants, dietary factors, stress and social factors. The review covers the effects of these exposures on various health outcomes, including cancer, developmental disorders and metabolic diseases. The authors highlight the importance of considering the role of the epigenome in environmental health research and the need for more interdisciplinary collaboration to develop effective prevention and intervention strategies. Overall, the article emphasises the critical role of the environment in shaping the epigenome and its impact on human health.

Prasher D, Greenway SC, Singh RB. The impact of epigenetics on cardiovascular disease. *Biochem Cell Biol.* 2020 Feb;98(1):12–22. doi: 10.1139/bcb-2019-0045. Epub 2019 May 21. PMID: 31112654.

This paper discusses the impact of epigenetics on cardiovascular disease. The authors provide an overview of the epigenetic modifications involved in cardiovascular disease, including DNA methylation, histone modification and non-coding RNAs. The article also reviews the impact of various environmental and lifestyle factors, such as diet, exercise and stress, on epigenetic regulation and cardiovascular disease risk. The authors discuss the potential for epigenetic therapies for cardiovascular disease treatment and prevention, as well as the challenges and limitations of such therapies. The article concludes by highlighting the need for further research in this field to improve our understanding of the complex interactions between epigenetics and cardiovascular disease.

Schrott R, Song A, Ladd-Acosta C. Epigenetics as a biomarker for early-life environmental exposure. *Curr Environ Health Rep.* 2022 Dec;9(4):604–24. doi: 10.1007/s40572-022-00373-5. Epub 2022 Jul 30. PMID: 35907133.

The article discusses the use of epigenetic modifications as biomarkers for early life environmental exposure. The authors provide an overview of epigenetics and how environmental exposures can alter epigenetic marks. They discuss the potential for these changes to serve as biomarkers for exposure to toxins, pollutants and other environmental stressors. The article also covers the challenges and limitations of using epigenetic biomarkers, including issues related to study design and data interpretation. The authors provide examples of how epigenetic biomarkers have been used to study the impact of environmental exposures on foetal development and childhood outcomes. The article concludes by highlighting the potential for epigenetic biomarkers to improve our understanding of the relationship between early life environmental exposure and long-term health outcomes. Overall, the article provides valuable insights into the use of epigenetics as a tool for environmental health research.

Tirthani E, Said MS, Rehman A. Genetics and Obesity. 2022 Aug 1. In: StatPearls [Internet]. Treasure Island (FL): StatPearls Publishing; 2022 Jan–. PMID: 34424641.

This paper discusses the role of genetics in obesity. The authors provide an overview of the genetic factors that contribute to obesity, including mutations in genes that regulate appetite and metabolism, as well as the complex interplay between genetics and environmental factors such as diet and physical activity. The article also reviews the diagnostic tools and treatment options for obesity, including genetic testing and personalised approaches to weight management. The authors conclude that a better understanding of the genetic basis of obesity could lead to the development of more effective prevention and treatment strategies for this increasingly common and serious health issue.

Zhang L, Lu Q, Chang C. Epigenetics in health and disease. *Adv Exp Med Biol.* 2020; 1253:3–55. doi: 10.1007/978-981-15-3449-2_1. PMID: 32445090.

This article provides a comprehensive overview of the role of epigenetics in health and disease. The authors discuss the distinct types of epigenetic modifications, including DNA methylation, histone modification and non-coding RNAs, and their influence on gene expression and cellular function. The article reviews the role of epigenetics in a range of diseases, including cancer, cardiovascular disease and neurological disorders, and discusses the potential of epigenetic therapies for disease treatment and prevention. The authors also highlight the importance of understanding the role of epigenetics in environmental exposures and lifestyle factors, such as diet and exercise, for disease prevention and personalised medicine. The article concludes with a discussion of the future of epigenetics research and its potential to transform our understanding of human health and disease.

6

Cancer Epigenetics

Cancer is one of the most common and severe diseases seen in clinical practice; early diagnosis and treatment are vital. The identification of persons at increased risk of cancer before its development is an important objective of cancer research. Cancer is not a single disease, but a term used to describe forms of neoplasia, a disease process characterised by uncontrolled cellular proliferation leading to a mass or tumour. Therefore, cancer is a disease of gene control, and given the integral role of epigenomics in gene control, studying cancer epigenetics is crucial to understand and better treat cancer.

For a neoplasm to be cancer, it also must be malignant, which means that its growth is no longer controlled and the tumour is capable of invading neighbouring tissues or spreading (metastasising) to more distant sites. Regardless of whether cancer occurs sporadically in an individual or repeatedly in many individuals in a family as a hereditary trait, cancer is fundamentally a genetic disease. Various kinds of genes have been implicated in initiating the cancer process, and diverse types of mutations are responsible for causing cancer. Once initiated, cancer evolves by accumulating additional genetic damage through mutations or epigenetic silencing of genes that encode cellular machinery that repairs damaged DNA and maintains cytogenetic normality. The biology of cancer is incredibly complex and a multistage process that can take an exceptionally long time from cancer initiation, through promotion to cancer progression and metastasis (Table 6.1).

Cancer can arise in any cell of the body and will present with two key dysfunctions: defective cellular proliferation and defective cellular differentiation. The genetic and epigenetic mutations that contribute to the progression of cancer from the initial initiation event enable the cancer to become increasingly independent from cell cycle control. There are numerous carcinogens and factors that can initiate cancer through cell/DNA damage; similarly, there exist multiple defence and repair mechanisms in the body that help to defend against cancer.

Epigenetics and Health: A Practical Guide, First Edition. Michelle McCulley.
© 2024 John Wiley & Sons, Inc. Published 2024 by John Wiley & Sons, Inc.

Table 6.1 Biology of cancer diagram.

DNA damage	DNA damage can be caused by a variety of factors, such as exposure to carcinogens, errors in DNA replication or inherited genetic mutations.
Mutation(s) in tumour suppressor or oncogene genes	Mutations in tumour suppressor genes, which normally help prevent uncontrolled cell growth and division, or oncogenes, which promote cell growth and division
Altered cellular signalling and proliferation	Can lead to changes in cellular signalling that allow for uncontrolled proliferation and survival.
Formation of a pre-cancerous lesion or tumour	The accumulation of mutations can lead to the formation of a pre-cancerous lesion or tumour, which is a clump of abnormal cells that can grow and divide uncontrollably
Tumour growth and invasion	
Metastasis to other parts of the body	The tumour can then invade nearby tissues and organs and potentially spread to other parts of the body through the bloodstream or lymphatic system.
Secondary tumour growth and invasion	Secondary tumours can then grow and invade other parts of the body, a process known as metastasis, which can be difficult to treat and often leads to poor outcomes.

Cancer manages to manipulate both the genome and the epigenome to achieve its end goal of escaping control and surveillance and becoming autonomous of the host. Because of the flexible nature of epigenetic mutations, it is likely that these are more used in cancer to evade immune surveillance and develop drug resistance. DNA methylation is somatically heritable and interacts with genetic changes to enable a cancer to evolve at a much faster pace than if it were dependent on genetic mutations alone. Research to date suggests a strong correlation between

Table 6.2 Approaching cancer from an epigenetic perspective.

Epigenetic change	Description	Example cancer types
DNA methylation	Hypermethylation of promoter regions can silence tumour suppressor genes, while hypomethylation of repetitive DNA elements can lead to chromosomal instability	Colorectal, lung, breast and other cancers
Histone modifications	Aberrant histone modifications can alter chromatin structure and gene expression, promoting oncogene activation or tumour suppressor gene silencing	Various cancers, including leukaemia, lymphoma and breast cancer
Non-coding RNAs	Altered expression of microRNAs, long non-coding RNAs and other non-coding RNAs can disrupt gene expression and cellular signalling, promoting cancer development and progression	Various cancers, including ovarian, lung and gastric cancer
Chromatin remodelling	Changes in chromatin remodelling complexes, such as SWI/SNF, can promote oncogene activation or tumour suppressor gene silencing	Paediatric cancers, including rhabdoid tumours and synovial sarcomas
DNA damage response	Disruption of DNA repair pathways, such as through mutations in BRCA1/2 or ATM, can promote genomic instability and cancer development	Breast, ovarian and other cancers with familial predisposition

disrupted epigenome and disruption to gene expression in cancer, making targeting and editing the epigenome a potential therapeutic approach to targeting aberrant gene expression in cancer cells. Table 6.2 summarises some of the roles of epigenetics in the development of cancer and is explained in more detail in the subsequent section.

6.1 DNA Methylation and Role in Cancer Development

Abnormal DNA methylation is a key feature of cancer; typically, this is manifested as genome-wide hypomethylation and genomic instability. Hypermethylation of CpG islands close to the transcription start sites of tumour suppressor genes results in their silencing; additionally, overexpression of proto-oncogenes is also commonly observed. Disruption of normal DNA methylation leads to cancer progression in both solid tumours and haematological malignancies; often, these aberrant DNA methylation patterns are associated with the loss of TET enzyme activity. Ten–eleven translocation family of enzymes has a role in catalysing the oxidation of 5mC to 5hmC and further products. Genes encoding the TET enzymes are frequently found to be mutated in human cancers.

In normal cells, DNA methylation serves several important functions, including the regulation of gene expression, maintenance of genome stability and suppression of repetitive elements. However, aberrant DNA methylation patterns are commonly observed in cancer cells, contributing to the initiation and progression of cancer. Hypermethylation of CpG islands located in gene promoter regions can lead to gene silencing. CpG islands are stretches of DNA rich in CpG dinucleotides that are often associated with gene regulatory regions. When CpG islands in promoter regions are methylated, the binding of transcription factors and other regulatory proteins is disrupted, preventing the initiation of transcription and leading to decreased gene expression. This can include tumour suppressor genes, which normally restrain cell growth and division. When these genes are silenced, it can contribute to uncontrolled cell growth and cancer development. DNA methylation-mediated silencing of tumour suppressor genes is a common event in cancer. Tumour suppressor genes, such as *p16INK4a*, *BRCA1* and *MLH1*, are critical in preventing uncontrolled cell growth and DNA damage repair. Methylation-associated silencing of these genes can remove crucial checkpoints that keep cell division in check, leading to increased genomic instability and cancer progression. DNA repair genes, responsible for fixing DNA damage and maintaining genome stability, can also be affected by DNA methylation changes. Methylation-induced silencing of DNA repair genes can result in an accumulation of mutations and genomic instability, both of which are hallmark features of cancer cells. Epithelial-mesenchymal transition (EMT) is a process by which epithelial cells lose their characteristics and acquire a more migratory, invasive phenotype. DNA methylation changes can contribute to EMT by influencing the expression of genes involved in cell adhesion, migration and invasion. This can promote the spread of cancer cells from the primary tumour to other tissues, leading to metastasis.

6.2 Histone Modification and Role in Cancer Development

Aberrant histone modifications have been strongly implicated in cancer development by affecting gene expression patterns and cellular behaviours. Histone acetylation involves the addition of an acetyl group to the lysine residues of histone proteins. This modification typically leads to an open chromatin structure, making the DNA more accessible to transcriptional machinery and promoting gene expression. In cancer, abnormal histone acetylation can lead to either hyperacetylation or hypoacetylation, both of which have implications for gene regulation. Hyperacetylation can occur in cancer due to the overactivity of histone acetyltransferases (HATs) or the loss of histone deacetylases (HDACs), which remove acetyl groups. Hyperacetylation can result in the activation of oncogenes that drive cell proliferation and survival. Conversely, hypoacetylation can lead to the silencing of tumour suppressor genes, which normally inhibit cell growth and promote apoptosis. HDAC inhibitors are a class of drugs developed to counteract this hypoacetylation and reactivate silenced genes.

Histone methylation involves the addition of methyl groups to lysine or arginine residues on histone tails. Unlike acetylation, the effects of histone methylation on gene expression can be context-dependent, depending on the specific residue being modified and the degree of methylation. For example, trimethylation of histone H3 at lysine 4 (H3K4me3) is associated with active transcription start sites of genes. Its misregulation can lead to inappropriate activation of oncogenes. Trimethylation of histone H3 at lysine 27 (H3K27me3) is associated with gene repression. Aberrant accumulation of this modification can lead to the silencing of tumour suppressor genes. Dimethylation or trimethylation of histone H3 at lysine 9 (H3K9me2/3) is generally associated with gene silencing and heterochromatin formation. Hypomethylation of H3K9 has been linked to the activation of oncogenes.

Phosphorylation of histones can affect chromatin structure and gene expression by altering the interactions between histones and other proteins. For instance, phosphorylation of histone H3 at serine 10 (H3S10ph) is associated with active transcription and is often found at the promoters of actively transcribed genes. Ubiquitination involves the addition of ubiquitin molecules to histone proteins. This modification can affect chromatin structure and gene expression. For example, monoubiquitination of histone H2B is associated with active transcription. Histone variants are slightly different forms of histone proteins that can influence chromatin structure and gene expression. Alterations in the expression or incorporation of histone variants, such as H2A.Z and H3.3, have been observed in cancer and can impact gene regulation. Histone modifications can also contribute to epigenetic memory in cancer cells. Once established, certain histone modifications can be faithfully inherited during cell division, perpetuating abnormal gene expression patterns that contribute to tumorigenesis.

6.3 Non-coding RNAs and Role in Cancer Development

Non-coding RNAs (ncRNAs) are a diverse group of RNA molecules that do not encode proteins but play critical roles in various cellular processes, including gene expression regulation. Over the past couple of decades, ncRNAs have emerged as key players in cancer development, progression and metastasis. They influence gene expression at multiple levels and contribute to the dysregulation of cellular processes in cancer cells. MicroRNAs are short (around 22 nucleotides) ncRNAs that post-transcriptionally regulate gene expression by binding to target mRNAs and promoting their degradation or inhibiting translation. In cancer, miRNAs are involved in both oncogenic and tumour-suppressive processes. Oncogenic miRNAs (OncomiRs) are often upregulated in cancer and promote tumourigenesis by inhibiting the expression of tumour suppressor genes. For example, miR-21 is frequently overexpressed in various cancers and targets several tumour suppressor genes, leading to enhanced cell proliferation, resistance to apoptosis and increased invasion and metastasis. Conversely, some miRNAs have tumour-suppressive functions by targeting oncogenes or genes involved in cell cycle progression. MiR-34a, a well-studied tumour suppressor miRNA, is downstream of p53 and promotes cell cycle arrest and apoptosis. Its downregulation is associated with numerous cancers. MiRNAs can also influence the metastatic potential of cancer cells. MiR-10b, for instance, promotes metastasis by targeting genes involved in cell adhesion and motility. Due to their pivotal role in cancer, miRNAs have gained attention as potential therapeutic targets. Modulating miRNA expression using synthetic miRNA mimics or inhibitors (antagomirs) holds promise for cancer treatment.

Long ncRNAs are RNA molecules longer than 200 nucleotides that do not encode proteins but perform various regulatory functions. They are involved in many aspects of cancer development and have been linked to processes such as proliferation, differentiation, metastasis and drug resistance. Numerous lncRNAs are upregulated in cancer and contribute to tumour growth. For example, *HOTAIR (HOX Transcript Antisense RNA)* is overexpressed in breast cancer and has been associated with increased invasiveness and metastasis. Certain lncRNAs act as tumour suppressors by regulating cell cycle progression, apoptosis and DNA repair. The lncRNA GAS5, for instance, inhibits cell growth by sequestering the glucocorticoid receptor and promoting apoptosis. Some lncRNAs influence gene expression by interacting with chromatin-modifying complexes, altering chromatin structure and affecting transcriptional regulation. LncRNAs like MALAT1 (Metastasis Associated Lung Adenocarcinoma Transcript 1) and HOTAIR have been implicated in promoting EMT and enhancing cell migration and invasion. The dysregulation of specific lncRNAs is associated with different cancer types and stages, making them potential biomarkers for cancer diagnosis and prognosis.

6.4 Chromatin Remodelling and Role in Cancer Development

Chromatin remodelling is a fundamental process that regulates the accessibility of DNA to various cellular machinery, including transcription factors, RNA polymerase and other regulatory proteins. It involves changes in the structure and organisation of chromatin, which consists of DNA wrapped around histone proteins. Chromatin remodelling is critical for controlling gene expression, DNA replication, DNA repair and other essential cellular processes. Dysregulation of chromatin remodelling has been strongly implicated in cancer development. Chromatin remodelling complexes are multi-protein complexes that use energy from ATP hydrolysis to alter the structure of nucleosomes (the basic units of chromatin) and modify chromatin accessibility. There are two main classes of chromatin remodelling complexes: ATP-dependent remodelers and histone-modifying complexes. ATP-dependent remodelers, such as SWI/SNF and ISWI, use the energy from ATP hydrolysis to slide, evict or change the position of nucleosomes along the DNA. This process can expose or hide DNA sequences, making them either more or less accessible to transcription factors and other regulatory proteins. Histone-modifying complexes, such as the Polycomb and Trithorax complexes, post-translationally modify histone proteins by adding or removing chemical groups like methyl, acetyl and ubiquitin. These modifications can influence chromatin structure and gene expression.

Dysregulation of chromatin remodelling is a hallmark of cancer development and progression. Changes in chromatin structure and accessibility can lead to aberrant gene expression patterns that contribute to various aspects of tumourigenesis. Chromatin remodelling can lead to the activation of oncogenes. Aberrant chromatin remodelling can create a permissive chromatin state that allows oncogenes to be expressed at higher levels, driving uncontrolled cell proliferation. Conversely, chromatin remodelling can lead to the silencing of tumour suppressor genes, which normally restrain cell growth and prevent cancer development. Inactivation of tumour suppressor genes through epigenetic mechanisms, including altered chromatin structure, is a common event in many cancers. Chromatin remodelling is crucial for DNA repair processes. Defects in chromatin remodelling can lead to compromised DNA repair mechanisms, accumulation of DNA damage and genomic instability, all of which are associated with cancer development.

6.5 DNA Damage Response and Role in Cancer Development

The DNA damage response (DDR) is a complex and highly regulated network of cellular processes that detects and repairs DNA damage, maintains genome stability and prevents the accumulation of mutations that can lead to cancer development. When DNA damage occurs, whether from endogenous sources (such as replication errors) or exogenous sources (such as radiation or chemical exposure), the DDR is activated to ensure proper repair or, in cases of irreparable damage, induce cell death (apoptosis). Dysfunction in the DDR can lead to genomic instability, a hallmark of cancer. Several types of DNA damage, including single-strand breaks, double-strand breaks and crosslinks, can trigger the DDR. Sensor proteins recognise the damaged DNA and initiate the signalling cascade. Ataxia Telangiectasia Mutated (ATM) and Ataxia Telangiectasia and Rad3-related (ATR) are key kinases involved in DNA damage sensing.

Once DNA damage is detected, a series of signalling events are initiated. The DDR involves several important pathways. The ATM pathway is activated by double-strand breaks; ATM phosphorylates key substrates involved in DNA repair and cell cycle arrest, such as p53 and checkpoint kinase 2 (Chk2). p53 activation can lead to cell cycle arrest and apoptosis, preventing the replication of damaged DNA. The ATR pathway is activated by single-strand breaks and stalled replication forks, ATR phosphorylates substrates such as Chk1, which leads to cell cycle arrest, allowing time for DNA repair.

The DDR induces cell cycle checkpoints that temporarily halt cell cycle progression to provide time for DNA repair. The G1/S checkpoint, S-phase checkpoint and G2/M checkpoint ensure that DNA replication and cell division do not occur until DNA damage is repaired. The DDR coordinates multiple DNA repair pathways to correct different types of DNA damage:

- Homologous Recombination (HR): Repairs double-strand breaks using the intact sister chromatid as a template.
- Non-Homologous End Joining (NHEJ): Repairs double-strand breaks by directly ligating broken DNA ends. It can be error-prone and may lead to mutations.
- Base Excision Repair (BER): Repairs damaged or mismatched bases by removing the damaged base and replacing it with the correct one.
- Nucleotide Excision Repair (NER): Removes bulky DNA lesions caused by UV radiation and certain chemicals.
- Mismatch Repair (MMR): Corrects mismatched bases and small insertion/deletion loops that arise during DNA replication.

A functional DDR is a critical component of tumour suppression. DDR defects can lead to the accumulation of mutations and chromosomal aberrations, contributing to genomic instability and the initiation of cancer. Inherited mutations in genes involved in the DDR, such as *BRCA1* and *BRCA2*, are strongly associated with an increased risk of developing certain cancers, including breast and ovarian cancer. Some cancer cells can exploit DDR pathways to their advantage. For example, they might have altered DDR components that allow them to tolerate higher levels of DNA damage, promoting survival and resistance to treatments like radiation and chemotherapy.

The relationship between the DDR and epigenetics is complex and interconnected. Epigenetic modifications can impact the expression of genes involved in the DDR. For example, changes in DNA methylation patterns can lead to the silencing of DNA repair genes, making cells more susceptible to accumulating DNA damage. DNA damage itself can lead to alterations in epigenetic marks. For instance, DNA damage can trigger changes in histone modifications and DNA methylation patterns at and around the site of damage. These changes can have long-term effects on gene expression and cellular function. Many cancers are characterised by widespread epigenetic changes. These changes can affect DNA repair genes and other genes involved in maintaining genomic stability. As a result, cells with compromised DDR due to epigenetic alterations are more likely to accumulate additional DNA damage, contributing to the cancer development process.

Understanding the interplay between DDR and epigenetics has implications for cancer therapy. Targeting epigenetic regulators can potentially sensitise cancer cells to DNA-damaging treatments like chemotherapy and radiation, making them more susceptible to cell death. Inhibiting DNA repair mechanisms in cancer cells that are already compromised in the DDR can enhance their sensitivity to treatments. PARP inhibitors, for instance, exploit deficiencies in HR repair and have shown promise in treating certain types of cancer.

6.6 Epigenetics and Metabolic Programming in Cancer

Another key feature of cancer is metabolic reprogramming, with cellular metabolism interacting with the epigenome and the molecular and genetic drivers that regulate cancer. In a cell, the metabolome and epigenome communicate bidirectionally, controlling metabolic reprogramming. Understanding this interaction alongside the interactions with both molecular drivers and epigenetic modifications in cancer is essential to the development of novel, effective cancer treatments. Many metabolic genes are regulated by epigenetic abnormalities, resulting in an aberrant rewiring of metabolic process, and redox homeostasis in cancer cells. Nutrients, including glucose, fatty acids and amino acids, are metabolised by cells, resulting in a range of metabolites, including acetyl-CoA, NAD+, SAM alpha-KG, ATP and succinate. These metabolites then function as cofactors or substrates that can modify proteins and chromatin. Such products of metabolism, through their alteration of chromatin structure, can then alter gene expression. This then supports the linkage between obesity as a risk factor for cancer. The mechanisms are complex, but an increase in body fat can thereby activate some mechanisms that could promote the initiation and progression of cancer. Studies have supported that high-fat, high-carbohydrate diets rich in specific elements such as antioxidants and phytoestrogens can modify the epigenome, and the ensuing modulation of gene expression can either promote or prevent cancer development. Dietary therapy, therefore, would be a convenient and cost-effective approach to cancer prevention and treatment; however, there is still insufficient research in this area before this can be progressed to clinical application.

6.7 Summary

The epigenetic profile or status of a cell has a vital role in determining the cell's differentiation pathway and function within an organism. Therefore, it makes sense that if these processes are disrupted, so too will the cells' capacity for normal cellular differentiation and functioning. Abnormal cellular differentiation and abnormal cellular functioning thus being characteristic features of malignancy. If a series of epigenetic modifications enable a cell to develop as part of differentiated tissue, then disruption of the epigenomic marks/mechanisms can therefore lead to a cell reverting to an undifferentiated state, which can potentially lead to malignancy.

As outlined in previous chapters, histone modifications, DNA methylation and noncoding RNAs are essential components in the maintenance of gene expression. Aberrant epigenetic regulation is associated with pathologies such as cancer. The potential reversibility of epigenetic marks makes them an attractive target for cancer research. Genome-wide maps comparing epigenetic modifications in normal and cancerous cells have demonstrated that there are specific epigenetic processes involved in both cancer initiation and progression. Typical dysfunctions include the dysregulation of epigenetic enzymes, resulting in abnormal gene expression, specifically of genes with a role in the cell cycle, including cell proliferation, cell differentiation and DNA repair. The cancer genome is globally hypomethylated, the exception being the promoter regions of several tumour suppressor genes, which are found to be hypermethylated, as expected given the role of DNA methylation in gene silencing. There is an extensive list of chromatin-controlling genes found mutated in cancer, elucidating how they precisely contribute to changing the epigenome and to cancer progression is the focus of many global research endeavours focused on the potential development of epigenetic therapies. The table below illustrates some such epigenomic mutations to illustrate the range of mutations, but it is by no means exhaustive.

Isocitrate dehydrogenase-encoding genes *IDH1* and *IDH2*	Gliomas, acute myeloid leukaemia	Inhibit histone demethylase and DNA demethylase	Altered DNA and histone methylation pattern
TET2	Myeloid malignancies	Hypermethylation of haematopoietic-specific enhancers	
DNMT3A	AML	DNA hypomethylation	
Histone H3K36	sarcoma	Dominant negative inhibition of methyltransferase	Global reprogramming of histone marks
Histone H3.3K27	gliomas	Dominant negative inhibition of methyltransferase	Global reprogramming of histone marks
H3K9ac, H3K18ac, H4K12ac			Global loss/low levels of acetylation of histones H3 and H4
H3K4me2 H4K20me3			Methylation of histones H3 and H4
lncRNA *HOTAIR*	Breast tumours and metastases	Regulate HoxD loci	Changes in histone post-transcriptional modifications mediated by histone modifier enzymes such as the polycomb repressive complex (PRC2), an H3K27 methylase.

The cancer genome atlas (TCGA) is a publicly available resource that has genomic, transcriptomic and proteomic data from over 30 cancer types and more than 20 000 primary cancers and matched normal samples. This is an incredible resource and an excellent starting place to explore cancer-related epigenomic data sets.

https://www.cancer.gov/about-nci/organization/ccg/research/structural-genomics/tcga

6.8 Epigenetic Alterations in Cancer and Therapeutic Design

Aberrant DNA methylation patterns can be detected in bodily fluids such as blood, urine and saliva, even before clinical symptoms of cancer appear. This has led to the development of DNA methylation-based biomarkers for early cancer detection and prognosis. These biomarkers can aid in identifying individuals at risk and monitoring the progression of the disease. The reversible nature of DNA methylation makes it an attractive target for cancer therapy. Drugs called DNA demethylating agents, such as 5-azacytidine and decitabine, can reverse DNA hypermethylation and reactivate silenced genes. These agents are used in certain cancer treatments, particularly for haematological malignancies. The epigenome is maintained and controlled by numerous proteins that control DNA methylation, histone modification and chromatin states. These proteins write, read and erase chemical marks from chromatin. In recent years, a significant amount of drug discovery work has been focused on this specific cellular machinery. Change in methylation pattern is particularly aligned with cancer initiation and progression. There are two characteristic methylation changes in cancer cells: firstly an overall decrease in global DNA methylation and secondly an increase in methylation around promoter CpG islands. Global hypomethylation found in cancer is most associated with the overexpression of proto-oncogenes and growth factors. Typically, hypomethylation is found at repeat sequences, CpG-poor promoters and retrotransposons. The activation of retrotransposable sites makes them active and contributes to genomic instability; similarly, with repeat sequences, instability arises from the lack of methylation. With respect to hypomethylation of growth factors, IGF-2 is a good example. It has a monoallelic expression pattern in non-malignant cells, but the loss of imprinting resulting from hypomethylation is found in the second allele of malignant cells, resulting in biallelic expression of the growth factor and uncontrolled tumour cell proliferation.

Interesting areas for further research are the mechanism/processes whereby some genes are hypomethylated in cancer, but others are spared. Key tumour suppressor genes/genes typically involved in growth suppression found hypermethylated and silenced malignancies including CDKN2A, p16, p73, p15 and TIMP-3. Additionally, as well as silencing of genes involved in cell growth, DNA methylation can contribute to cancer initiation/progression indirectly. For example, hypermethylation of transcription factors (such as RUNX3, GATA-4 and GATA-5) and DNA repair genes (such as BRCA1, MGMT, MLH1, MSH2 and ERCC). Analysis of DNA methylation in such genes is increasingly used as prognostic biomarkers for cancer survival and treatment response.

Targeting chromatin remodelling components has emerged as a potential strategy for cancer therapy. Small molecules that inhibit specific chromatin remodelling enzymes or complexes are being explored as potential treatments. For instance, inhibitors of HDACs and enhancer of zeste homolog 2 (EZH2), a subunit of the polycomb repressive complex 2 (PRC2), are being studied for their potential to reverse abnormal chromatin states in cancer cells.

Changes in histone proteins in cancer are found globally across genomic DNA and at specific genetic loci. Taking acetylation as an example, the inclusion of an acetyl group at a lysine residue of a histone tail can change chromatin compaction and regulate intracellular pH. The association between low pH and decreased histone acetylation has been found in many cancers. Additionally, changes to global levels of histone acetylation have also been linked to a range of cancers. HATs facilitate the incorporation of acetyl groups into the lysine tail of a histone protein, hyperacetylation arises from overactivity of the enzyme, resulting in hyperacetylation and activation of proto-oncogenes. Tumour suppressor genes become inactivated because of hypoacetylation. The balance between histone acetylation and deacetylation is disrupted in cancer cells.

Most miRNAs have a role in a process relating to cell cycle/growth, as such alterations in the regulation of miRNAs predominately via epigenetic mechanisms are implicated in cancer pathogenesis. miRNAs miR-148a, miR-148b and miR-152 have all been identified as having a significant role in the development of cancer and potential oncogenes/tumour suppressor genes. Other miRNAs specifically identified with malignancies include miR-15 and miR-16, miR-9, miR-34a, miR-137, miR-124 and miR-200.

6.9 Conclusion

The dearth of genetic mutations in epigenetic regulatory complexes means that there are several crucial targets for anti-cancer epigenetic drug discovery; choosing the right target and drug is still problematic. There are so many options to target and bringing a drug to market is extremely costly so, most of the research is focused on epigenetic targets, which are enzymes that can erase epigenetic marks, read epigenetic marks and target the chromatin complex. Further molecular characterisation of cancers to identify clinically relevant biomarkers will be a priority in cancer drug discovery. A big problem with epigenetic drugs is the lack of selectivity of the drugs, e.g. HDAC inhibitors are non-selective towards HDAC enzyme meaning both desired and non-desired regions are targeted contributing to undesirable side effects. The most widely studied and understood histone modifications are methylation and acetylation, both can be therapeutically targeted. Over the last few years, there has been a rapid increase in research into the development of epidrugs to specifically target cancer-related aberrant epigenetic changes. In summary, epidrugs can be designed to focus on the following modes of action:

a) alter DNA methylation
   a) DNMT inhibition
      i) Nucleoside analogues, e.g. Azacytadine, Decitabine, Zebularine
      ii) Non-nucleoside analogues e.g. polyphenols

b) alter histone modification
   a) histone methyltransferase inhibitors
   b) histone demethylase inhibitors
   c) histone acetyltransferase inhibitors
   d) histone deacetylase inhibitors

There are now many clinical trials using inhibitors of epigenetic mechanisms such as HDAC inhibitors and DNMT inhibitors. The mode of action of these drugs is to remove aberrant methylation, inhibiting DNMTS and inducing genome reprogramming. Some epidrugs are currently used alone or alongside other cancer therapies, the main challenge remaining is the reduction of cytotoxicity and side effects to maximise the positive outcome for the patients.

Task

Cancer is a group of heterogeneous conditions. Identify a specific cancer and approach it from the following two perspectives:

- Understanding its ontogeny from an epigentic perspective
  Here we are focusing on understanding the cancer to aid in cancer prevention, are there environmental and genetic risk factors for the cancer in question? How does epigenetics play a part in the progression of the cancer?
- Development of an epidrug to suppress the cancer growth
  Here we are focusing on a solution to try to cure the cancer in question. What are the key proteins that are typically overexpressed in the cancer? Can these be targeted and supressed through epigenetic control?

Further Reading

Feinberg AP, Levchenko A. Epigenetics as a mediator of plasticity in cancer. *Science.* 2023 Feb 10;379(6632):eaaw3835. doi: 10.1126/science.aaw3835. Epub 2023 Feb 10. PMID: 36758093.

This paper discusses the role of epigenetics in the development and progression of cancer. The authors argue that epigenetic changes play a crucial role in mediating the plasticity of cancer cells, enabling them to adapt and evolve in response to environmental pressures such as chemotherapy and immunotherapy. The article reviews recent research on epigenetic modifications in cancer cells and highlights the potential of epigenetic therapies for cancer treatment. The authors conclude that a better understanding of the role of epigenetics in cancer could lead to the development of more effective and personalised treatments for the disease.

Kanwal R, Gupta K, Gupta S. Cancer epigenetics: an introduction. *Methods Mol Biol.* 2015;1238:3–25. doi: 10.1007/978-1-4939-1804-1_1. PMID: 25421652.

This paper introduces cancer epigenetics. The authors discuss the various epigenetic modifications involved in cancer development and progression, including DNA methylation, histone modification and ncRNAs. The article also reviews the impact of environmental factors, such as tobacco smoke and radiation, on epigenetic regulation and cancer risk. The authors discuss the potential for epigenetic therapies for cancer treatment and prevention, as well as the challenges and limitations of such therapies. The article concludes by highlighting the need for further research in this field to improve our understanding of the complex interactions between epigenetics and cancer.

Recillas-Targa F. Cancer epigenetics: an overview. *Arch Med Res* 2022 Dec;53(8):732–40. doi: 10.1016/j.arcmed.2022.11.003. Epub 2022 Nov 18. PMID: 36411173.

The article provides an overview of cancer epigenetics. The author describes the role of epigenetic changes in cancer initiation, progression and response to therapy, and provides examples of how specific epigenetic alterations contribute to different cancer types. The article also highlights the potential of epigenetic drugs as cancer therapies, as well as the challenges and limitations in targeting epigenetic mechanisms for cancer treatment. The author emphasises the importance of continued research in cancer epigenetics to improve our understanding of cancer biology and develop more effective and personalised cancer treatments.

Sapienza C, Issa JP. Diet, nutrition, and cancer epigenetics. *Annu Rev Nutr* 2016 Jul 17;36:665–81. doi: 10.1146/annurev-nutr-121415-112634. Epub 2016 Mar 23. PMID: 27022771.

This article discusses the relationship between diet, nutrition and cancer epigenetics. The authors provide an overview of the key epigenetic modifications involved in cancer development and progression. The article also reviews the impact of dietary factors such as folate, micronutrients and bioactive compounds on epigenetic regulation and cancer risk. The authors discuss the potential for dietary interventions to modulate epigenetic processes and prevent cancer, as well as the challenges and limitations of such interventions. The article concludes by highlighting the need for further research in this field to improve our understanding of the complex interactions between diet, epigenetics and cancer.

7

Mental Health Epigenetics

In this chapter, we are going to discuss epigenetic health, its role in mental wellbeing and the neuroepigenome. Psychiatric disorders are a heterogeneous group of complex disorders that are the result of a combination of genetic and environmental factors and the interplay between them. In this chapter, we discuss transgenerational epigenetic inheritance, specifically in the context of psychiatric conditions. Because epigenetic control is susceptible to environmental interactions and also has the potential for reversal, this poses an interesting and major area in epigenetic research. As with many other complex disorders, the application of epigenomics to psychiatry has key areas of potential application:

- Epigenomic changes as diagnostic markers of disease phenotype
- Epigenomic changes as markers of disease progression/response to treatment
- Design of pharmaceuticals specifically targeting epigenomic changes associated with mental health disorders

The WHO constitution states, 'health is a state of complete physical, mental and social well-being and not merely the absence of disease or infirmity'. Mental health is a state of well-being in which an individual realises their own abilities, can cope with the normal stresses of life, can work productively and is able to contribute to their community. More than 100 million people in the Western Pacific Region (which includes New Zealand) suffer from mental health challenges. Poor mental health is associated with multiple social, biological and psychological factors, for example, persistent socioeconomic pressures, poverty, low levels of education, rapid social change, stressful working conditions, gender discrimination, social exclusion, unhealthy lifestyle, risks of violence and physical ill health, human rights violations as well as personality factors, genetic factors and chemical imbalances in the brain.

Increasing evidence demonstrates that the methylome changes significantly throughout the human lifespan, with documented influences including environmental, dietary, lifestyle and physical factors as well as more recent

Epigenetics and Health: A Practical Guide, First Edition. Michelle McCulley.
© 2024 John Wiley & Sons, Inc. Published 2024 by John Wiley & Sons, Inc.

Table 7.1 Epigenomics and psychiatry overview.

Psychiatric condition	Epigenetic contribution
Autism spectrum disorder	DNA methylation changes in genes involved in neurodevelopment, immune function and synaptic plasticity.
Schizophrenia	Aberrant DNA methylation and histone modifications in genes related to neurodevelopment, neurotransmitter signalling and synaptic function.
Bipolar disorder	Altered DNA methylation in genes involved in stress response, neurotransmitter signalling and circadian rhythms.
Major depressive disorder	Changes in DNA methylation and histone modifications in genes associated with stress response, neuroplasticity and inflammation.
Post-traumatic stress disorder	Altered DNA methylation and histone modifications in genes involved in stress response, neuroplasticity and immune function.
Addiction	Changes in DNA methylation and histone modifications in genes related to reward pathways, stress response and synaptic plasticity.
Obsessive-compulsive disorder	Altered DNA methylation and histone modifications in genes involved in neurodevelopment, neurotransmitter signalling and synaptic function.
Attention deficit hyperactivity disorder	DNA methylation changes in genes related to dopaminergic signalling and neural development.

evidence that psychological exposures also have a significant impact on the methylome. There are many animal studies providing evidence that anxiety, stress and ill-treatment can have a significant impact on methylome in rats and macaques. Human studies investigating the relationship between DNA methylation and psychological distress have found evidence that in humans, anxiety, depression, stress, neglect, violence, maternal care and socio-economic status all impact DNA methylation levels. The aberrant methylation patterns associated with psychological distress in humans are not restricted to changes in promoter methylation regions but also include changes to methylation within the gene body, intergenic and untranslated regions (Table 7.1).

7.1 Specific Genes of Interest with Regards to Mental Health

This area is very much in the discovery phase, but there are four genes that are often associated with psychological illness and have been the most extensively researched thus far. These genes are *NR3C1, SLC6A4, BDNF* and *OXTR* and have been widely associated with depression, stress, anxiety, neglect and abuse in humans. DNA methylation patterns fluctuate from the point of fertilisation through childhood, adolescence and adulthood, and as such, studies explore the impact of psychological trauma on the methylation of these genes at different life points.

7.1.1   *NR3C1* – Glucocorticoid Nuclear Receptor Variant 1

In animal models, differences in maternal care impact on NGF1-A binding site in the rat Nr3c1 gene. The binding site has a role in controlling hippocampal glucocorticoid receptor expression. Prenatal exposure to stress then leads to an increased corticosterone response to minor stressors in adulthood, reducing the expression of the glucocorticoid receptor. The HPA system is regulated by the glucocorticoid receptor, and as such is sensitive to the effects of adverse early life experiences and can therefore be impacted by maternal stress hormones released in the developing foetus. A human study testing this found that maternal depression in the third trimester impacts methylation of the *NR3C1* gene and in turn, affects the infant stress response (Oberlander et al. 2008), another study found that offspring of victims of domestic violence had increased methylation in the GR promoter, the children tested were aged between 10 and 19, illustrating the persistence of the disrupted DNA transgenerational methylation pattern (Radtke et al. 2011). Suicide completers who had traumatic childhoods were found to have higher methylation in the *NR3C1* promoter region than completers with no history of traumatic childhoods. There are many other similar studies, suggesting that the methylation of the *NR3C1* promoter could be a useful marker for identifying those who are at risk of developing severe psychopathological symptoms later in life.

In rats, because of early life maltreatment, the epigenetic status of Nr3c1 in the hippocampus is associated with the quality of parental care, levels of glucocorticoid receptor and vulnerability to stress. In humans, there are DNA methylation changes in the *NR3C1* gene in the hippocampus of suicide victims with a history of childhood abuse compared to suicide victims/controls with no reported abuse. Many other studies looking at *NR3C1* in association with a range of psychiatric conditions consistently detect DNA methylation changes. The amplitude of methylation changes at the glucocorticoid receptor gene in adults with psychiatric conditions is correlated with the severity/repetition of maltreatment in childhood. Altered glucocorticoid receptor levels induced by DNA methylation impair negative feedback of the HPA axis; hence, the stress responses become inadequately regulated.

Other key genes in the HPA-regulated stress response are the *FKBP5* gene (regulates GR), hypothalamic arginine vasopressin (*AVP*), corticotropin-releasing hormone (*CRH*) and pituitary melanocortin (*POMC*) genes. Similar genetic-epigenetic interplay has been demonstrated for the regulation of *FKBP5*, specifically with regards to early life trauma, whereby exacerbated activation of an *FKPP5* risk variant results in the active demethylation of a distal regulatory locus; the subsequent hypomethylation results in even greater *FKBP5* expression, further sensitising GR to future exposure. Disinhibition of the sensitivity of GR and impaired HPA axis results in an aggravated and prolonged stress response, a change in stress-related circuits, synaptic plasticity and emotion-processing brain structures that ultimately affect stress coping and vulnerability to psychiatric disorders.

7.1.2   *SLC6A4* – Serotonin Transporter

The serotonin system mediates susceptibility to stress; therefore, it follows that epigenetic marks may be involved in the regulation of serotonin through alteration of transcription of the serotonin transporter gene. The serotonin transporter removes the 5HT released in the synaptic cleft. Altered DNA methylation of the *SLC6A4* gene may explain how adverse events might lead to less than optimal socio-emotional stress regulation and further in life health and disease. In human studies, higher *SLC6A4* promoter region methylation has been associated with reduced mRNA levels in cell lines. Hypermethylation at specific CpG sites located within *SLC6A4* exon 1 has been found in studies looking at the impact of early traumas in childhood. Although there is evidence pointing towards hypermethylation of *SLC6A4* being a marker of early adversity exposure, more work still needs to be done to resolve issues with respect to tissue-specific expression. Genetic polymorphism-related differences in *5-HTTLPR* have an impact on *SLC6A4* methylation, as well as study design and population group limitations (ethnicity, gender, size, time point, etc.)

The link between aggression and brain dysfunction points towards neurochemical systems for explanations; specifically, low levels of serotonin have been associated with impulsive aggression and a lifetime risk for depression. DNA methylation of the promoter of the serotonin transporter gene controls serotonin mRNA levels; this is variable with genotype in the 5-HTTLPR gene. SLC6A4 methylation has been looked at in association with a range of prenatal and postnatal adverse exposures and is a confirmed biomarker of early adversity exposure. Epigenetic marks on this gene are believed to have a critical role in programming long-term health and disease. Looking at methylation of *SLC6A4* in cohorts genotyped for polymorphism in *5-HTTLPR* – numerous studies reported moderator effect of the *5-HTTLPR*-polymorphism on *SLC6A4* gene DNA methylation in association with stressful life events (childhood abuse, neighbourhood crime or socioeconomic status) and with psychopathological risk (depression, antisocial personality disorder, antidepressant response). In some cases, only S allele carriers were found to be epigenetically susceptible to stressful life events; in other cases, only LL homozygotes showed a correlation between DNA methylation changes and exposure to stress.

7.1.3   *BDNF* – Brain-Derived Neurotrophic Factor

Brain-derived neurotrophic factor (BDNF) is a mediator of neural function and plasticity. It is a neurotrophin with a key role in the growth and development of neurons as well as normal maturation of the neural development pathways. Studies in children have found specific methylation profiles of BDNF associated with depression, post-traumatic stress disorder (PTSD) and trauma-related disorders. In adults, BDNF is important for dendritic growth, synaptic plasticity, learning, long-term memory with disruptions to gene activity associated with depression, varied responses to social stress, anxiety, drug addiction and aggressiveness. *BDNF* encodes one of the most prevalent brain growth factors, and it is known to have a key role in the development, plasticity and survival of several neuronal subtypes. Studies in rats have shown that stress-induced depression and PTSD result in long-lasting repressive epigenetic marks of Bdnf in the brain, lower levels of Bdnf expression and depressive-like behaviour. In studies in humans, on peripheral blood, higher levels of methylation of *BDNF* were found in humans exposed to low childhood maternal care compared to those with high maternal care. Similarly, repressive hypermethylation of *BDNF* was also found in depressive patients. These changes are also associated with a previous history of suicidal attempts, ideation and poor response to antidepressant treatment.

7.1.4   *OXTR* – Oxytocin Receptor

Oxytocin is a neuropeptide widely expressed in the central nervous system. It has been reported to have a vital role in social behaviours and interactions, including trust, empathy, envy and attachment. Genetic variants in the oxytocin receptor (*OXTR*) gene impact social behaviour, with particular genotypes having low empathy, low optimism and low self-esteem, as well as higher states of loneliness. Variable DNA methylation patterns in the *OXTR* gene and subsequent changes in gene expression have been implicated in social/behavioural childhood disorders.

7.2   Specifically Focussing on Schizophrenia

There are three core pathogenic pathways crucial in the development of schizophrenia: abnormal brain development, impaired synaptic plasticity and altered glutamatergic function. Like all complex conditions, understanding the interaction between genetic and environmental factors is crucial. Studies of schizophrenia have helped further understanding of the dynamic qualities of the epigenome and its role in altering gene regulation through brain development. In terms of the genetic factors that have been identified as associated with the development of schizophrenia, the majority (more than 90%) are located in non-coding regions of the human genome, which suggests that schizophrenia likely develops from regulatory disruptions. Regulatory sequences are often the regions of the genome found to be methylated; histones are also subject to epigenetic modifications associated with either activation or inhibition of gene transcription. A plausible hypothesis for the role of epigenetics in the pathogenesis of schizophrenia is that a combination of epigenetic and genetic risk factors alters key gene regulation during early brain development processes, priming the brain at a risk state for schizophrenia.

According to current research investigating the epigenetics of specific cell types in schizophrenia, key alterations are present in glutamatergic neurons in association with schizophrenia. 3D genome studies that look at the interactory nature of the genome have linked specific schizophrenia genetic variants to distal enhancer elements that mostly interact with glutamatergic genes. These data suggest a potential contributing factor in the molecular ontogeny of schizophrenia is the disruption of chromatin loops that activate gene expression in glutamatergic neurons. What interactions are occurring because of genomic and epigenomic variants? To answer this question, we can integrate a 3D genome with regulatory histone modification maps of specific neuronal subtypes to further understand the gene regulation networks in certain cell types that drive the development of psychiatric conditions.

Even given the rapid development of techniques for studying dynamic epigenomics and the interaction with genomics, we still only have a rudimentary understanding of the molecular pathology leading to schizophrenia. There are many challenges, of course, and hopefully with developments such as PsychENCODE and the advent of advancements in technology such as 3D cell culture/organoids and spatial omics, there will be greater integration of epigenomic, genomic and phenomic data.

7.3 Transgenerational Epigenetic Influences on Predisposition to Psychiatric Disorders

There are evolutionary advantages to the transgenerational transmission of epigenetic marks. It is a rapid way to pass on adaptation to new environmental encounters, a way of transferring an adaptive response. Early life adversity modifies the epigenome of the HPA axis and is associated with a wide range of mental health disorders. Effects of environmental stimuli during pregnancy, parental care, adulthood and germline transmission have all been proposed as precursors of epigenetic change that can be inherited transgenerationally. There is evidence to suggest that there is a transgenerational impact of both negative and positive environments on behavioural phenotypes.

Offspring of trauma survivors are at an increased risk for mental health problems; children of holocaust survivors have an increased risk for PTSD, an elevated risk of anxiety and depression, impaired cortisol levels and modified epigenetic regulation of *NRC31*, suggesting a role of the HPA axis in this type of transgenerational inheritance. Studies have been carried out to investigate the impact of paternal stress on affective behaviours and HPA-axis regulation of their progeny by increasing levels of the stress hormone corticosterone. A mouse model of chronic social defeat stress demonstrated that the offspring of stressed adult male mice have increased affective behavioural responses and increased plasma CORT and VEGF levels in their male offspring, alongside dysregulation of the HPA axis. Paternal experience was also found to affect offspring cognitive abilities, with paternal exposure to corticosterone influencing offspring cognitive performance. The female offspring only of CORT-treated fathers display reduced memory retention. Most mammalian traits are influenced by gender; the differences in stress response can potentially explain the sex-dependent inheritance observed in animal studies of paternal stress. Paternal depression is associated with an increased risk of depression in adolescent boys compared to girls.

In contrast, stress-positive environmental factors can improve the stress response and cognitive abilities in both rodent models and humans. A positive environment could include enhanced physical exercise, environmental enrichment or other forms of enhanced motor, sensory and cognitive stimulation. Environmental enrichment can reverse the effects of adverse life events, possibly through epigenetic effects on the DNA methylation profile. Three months of exercise training in humans were found to modify the methylation profile in sperm, exposing obese male mice to swimming as an exercise that normalised their sperm miRNA profile. Also in mice, increased paternal physical activity suppressed juvenile fear memory and reduced anxiety-like behaviour, possibly via changes in levels of specific sperm small non-coding RNAs. Combined, this evidence suggests physical activity in fathers has the potential to alter offspring behaviour traits via epigenetic modifications in the paternal sperm.

Environmental enrichment has been shown to rescue the transgenerational influence of several factors. The beneficial properties of environmental enrichment in ameliorating adverse exposure can also be transmitted to the offspring, and this has important public health implications for humans. By 'removing' the epigenetic memory of stressful life events in fathers at elevated risk, we could rescue the epigenetic alterations in the sperm and prevent detrimental changes to the offspring. E.g. therapeutic interventions before conception in subjects who have experienced severe stress and a high predisposition for psychiatric disorders. Environmental enrichment is synergistic to medication; in mice, the beneficial impact of SSRIs has the most effect in an enriched environment. Additionally, SSRIs induce upregulation of several miRNAs, including mi135 and miR-375. miR-135 is decreased in the blood of depressed patients and in the brains of depressed suicide victims; it is increased after cognitive behavioural therapy in depressed patients, suggesting a similar effect of SSRIs and behavioural treatment.

7.4 Suicide/PTSD

As parents, we do not only transmit our genes to our offspring but also imprints of significant traumatic experiences. As discussed above, epigenetic mechanisms provide an adaptive response to novel, unpredictable environmental change. Evidence suggests such an imprint can be transmitted to subsequent generations, increasing their susceptibility and vulnerability to mental health issues. PTSD is a psychological reaction to experiencing or witnessing a significantly stressful, traumatic or shocking event. Both genetic and epigenetic factors contribute to PTSD risk. There are many convincing research studies suggesting epigenetic mechanisms play a considerable role in stress regulation and contribute significantly to the development of PTSD. It can be hypothesised that maladapted phenotypic plasticity, mediated by epigenetic mechanisms, produces an epigenetic signature associated with PTSD. This imprint, by its nature, has the potential to be transmitted to offspring but also has the potential to be reversed with appropriate pharmacotherapy and psychosocial intervention. The signature can be used to monitor recovery progress. The identification of a reversible molecular imprint can serve to normalise PTSD and the associated debilitating changes in mental health. Monitoring levels of a PTSD biomarker, in the way diabetics monitor insulin levels, could help sufferers monitor their mental health and ensure they are appropriately supported both through pharmacological and psychosocial intervention to achieve posttraumatic growth and recovery.

7.5 Stress, Epigenetics and Transgenerational Epigenetic Inheritance: Consequences of Inequity/Deprivation

Can the stressors we are exposed to have a legacy effect on subsequent generations in terms of the epigenomic imprint transmitted to progeny? If so, this has profound implications in terms of how we assess the heritability of disease, health equity and disease risk. The exquisite interplay between genome and environment is what constitutes epigenetics. Unlike our DNA sequences, which are largely the same in every cell, epigenetic changes can occur because of dietary and other environmental exposures, including physical and psychological stressors. The epigenome can be viewed as an important modifier of disease susceptibility, and as such, is particularly vulnerable to environmental interference at particular life cycle points, in particular, periods of rapid growth and change such as embryogenesis, neonatal development and puberty. Abnormalities in DNA methylation are associated with many diseases, including cancer, and manifest through inappropriate gene expression. As discussed prior, many environmental toxins, including those associated with pesticides, are known to produce toxic effects resembling methyl insufficiency, reducing gene inhibition and increasing gene expression of sensitive genes.

Many epigenetic changes have been found to be inherited trans-generationally in experimental models; however, little is known as to whether environmentally induced changes in epigenomic gene control persist in successive generations and if so, how long for; this is a fundamentally crucial question to answer and could have profound implications for addressing health equity and disease risk in human populations. If your grandparents are exposed to chronic stressors, how many generations does this epigenomic imprint carry through, and what are the longer-term consequences for population health? If epigenetic inheritance occurs in humans, then environmental risk factors, particularly prevalent in adversity may epigenetically impact both the individual, their children, and their grandchildren. Epigenetics potentially affects factors associated with disadvantage, including stress response, emotional regulation, cognitive development, disease susceptibility and mental disorders. To date, most studies have not demonstrated causation, and research is needed as to the genetic mechanisms underlying this transmission. Animal models suggest that both maternal and paternal germlines can be induced by environmental stress and can be passed on to subsequent generations.

This then leads to the intergenerational transmission of socioeconomic disadvantage and the developmental origins of adult health and disease. Whereby foetal exposure to maternal prenatal distress associated with socioeconomic disadvantage compromises the offspring. Particularly via the HPA axis regulation. Recent evidence suggests the effects of a broader timeframe could manifest in offspring, with parents' childhood experiences having the ability to transfer epigenetic marks that could impact the development of their offspring independently of perinatal environment and early childhood environment relating to socioeconomic disadvantage. The difficulty in human populations is being able to carry out our longitudinal studies, but if research does support this, then this has important policy implications in terms of reducing the continuation of disadvantage across generations, which needs to be addressed to understand the perpetuation of compromise and its impact upon health across multiple generations.

Household and neighbourhood poverty are related to stress and the cumulative effect of stressors (housing/food insecurity, child abuse, neglect, substance abuse, violence) can induce a toxic stress response in young children and lead to long-term changes in brain structure and function. Poor nutrition and obesity contribute to chronic physical and mental health problems across the life course. Lack of access to adequate healthcare, smoking and drug use and lack of physical activity worsen problems. Low birth weight is used as a predictor of future disease risk. Foetal programming/developmental origins of health and disease (DOHaD) model illustrates how foetal exposure to a variety of maternal life experiences (poor nutrition, pollutants, stress) can affect an offspring's neurodevelopment and have implications for future health and wellbeing. There is compelling evidence for mother-to-child influence occurring prenatally, in part through the HPA axis regulation.

Poor genes do not cause poverty; adversity affects growth and development, limits/prematurely restricts individuals' productive potential. Epigenetics suggests there is an expanded timeframe for intergenerational impact outside of just the perinatal period. Early life stress in the mother appears to alter the placental environment regardless of a woman's health and experiences during pregnancy, causing de novo epigenetic changes in the developing embryo that could mimic the changes identified in the mother and affect foetal development. For example, maternal early life adversity might alter egg cell cytoplasm that after conception exerts an influence on the developing foetus. Another scenario is that increased placental concentrations of endocrine and immune stress mediators (resulting from early life adversity in the mother) could cause alterations in the expression of miRNA and DNA methylation in the foetal brain, which could in turn bring about changes in foetal cell proliferation, neuronal differentiation, gliogenesis, availability of neurotrophic growth factors, cell survival synaptogenesis, neuro-transmitter levels, myelination and adult neurogenesis. Support for this hypothesis comes from evidence that adversity in early childhood produces long-term alteration in endocrine and immune-inflammatory physiology, including greater HPA axis reactivity and greater pro-inflammatory state.

Key questions that need to be addressed in research are: what are the mechanisms by which this occurs, and how can epigenetic reprogramming be avoided?

Intrauterine exposures, mutations in DNA repair mechanisms, shared environments and reproduction in subsequent generations of the environment/behaviours that influenced gene expression in the first generation are all likely to contribute. No robust evidence really exists for how epigenetic changes in germ cells escape the reprogramming that occurs after fertilisation and as the embryo and foetus are developing. Differences between humans and mice do not permit direct inference to humans, and additionally, animal studies cannot replicate the complex social and community structures. Therefore, there is a need for human cohort studies to provide unambiguous evidence of intergenerational epigenetic transmission in humans.

Through comparative analysis of methylome data, we can attain a clearer understanding of the molecular mechanisms that underlie and link environment and genome in terms of environmental exposure to toxins and the phenotypic outcome of health and recovery of the individual and the population. Heat shock Protein 90 (HSP90) has been identified, and well established as a major controller of cell growth and proliferation associated with cancer (Khurana and Bhattacharyya 2015). Previous studies have shown that even mild environmental stressors can have a role in gene silencing of the heat shock protein HSP90. HSP90 is inhibited by anti-tumour antibiotics, and this inhibition has been shown to be transmitted to subsequent generations; therefore, it is plausible to expect other stressors to have similar effects on epigenetic machinery. This buffering capability of certain genes, engaging in a molecular homeostasis function through epigenetic mechanisms is conserved across the animal kingdom, when this capacity for buffering is disrupted by stressors such as temperature and chemical stressors, cryptic phenotypic variants are seen to be expressed even after the gene function is restored. This has profound implications for how we assess heritable risk, with genetic variation no longer being the major determining factor in health risk, what is more important is the ability to buffer environmental/life stress.

From human studies, children show an increased risk for asthma if their fathers, but not their mothers, smoked before conception, when smoking occurred at early puberty, suggesting that this is a critically vulnerable period of the sperm. Several loci that resist DNA methylation reprogramming, including a number of loci associated with neurological disorders, are thus potential candidates for transgenerational epigenetic inheritance. DNA methylation profile of well-known obesity-related genes varies between sperm of obese and lean males. Bariatric surgery in obese men reversed the epigenetic profile in DNA loci that regulate appetite, demonstrating a direct response to the sperm content due to the surgery-induced change in metabolism.

7.6 Conclusion

This is a fascinating, incredibly complex and uniquely human field. The take-home message is that life experiences, whether stressful or positive, can alter molecular signals in the sperm and affect an offspring's physiological and behavioural phenotypes in a gender-specific manner. It is possible that these transgenerational epigenetic mechanisms may have evolved to help offspring better adapt to the changing environment into which they are born. Calibrating offspring phenotypes to dynamic and stochastic environments; together genomes and enviromes combine to provide a unifying explanation of health and disease processes as well as suggest novel therapeutic approaches for improving future human health. Understanding how parental lifestyle, experience and environmental exposures impact on health and disease in offspring should be an urgent priority and facilitate the potential development of novel approaches to prevent increased risk for specific disorders in subsequent generations. Several loci escape DNA methylation reprogramming in human germ cells and are associated with neurological disorders such as schizophrenia and multiple sclerosis. These resistant loci can potentially transfer epigenetic information to subsequent generations. A better understanding of the role of transgenerational inheritance in humans can provide us with the opportunity to predict, prevent and treat ill health, particularly psychiatric disorders, using a personalised/precision medicine approach.

Task

- *What are some of the broader consequences of your research questions?*
- *What are the key issues that need to be addressed?*

References

Khurana N, Bhattacharyya S. Hsp90, the concertmaster: tuning transcription. *Front Oncol.* 2015 Apr 28;5:100. doi: 10.3389/fonc.2015.00100. PMID: 25973397; PMCID: PMC4412016.

This article discusses the role of heat shock protein 90 (Hsp90) in regulating gene transcription. The authors describe how Hsp90 interacts with multiple transcription factors and co-chaperones to regulate their stability, activity and chromatin binding. The article also highlights the potential of Hsp90 as a therapeutic target for cancer, as it plays a critical role in the function of many oncogenic transcription factors. The authors conclude that Hsp90 acts as a 'concertmaster' in tuning the transcriptional orchestra and suggest that targeting Hsp90 may provide a novel therapeutic approach for cancer treatment.

Oberlander TF, Weinberg J, Papsdorf M, Grunau R, Misri S, Devlin AM. Prenatal exposure to maternal depression, neonatal methylation of human glucocorticoid receptor gene (NR3C1) and infant cortisol stress responses. *Epigenetics.* 2008 Mar–Apr;3(2):97–106. doi: 10.4161/epi.3.2.6034. PMID: 18536531.

The study examined the relationship between prenatal exposure to maternal depression, methylation of the NR3C1 gene and infant cortisol stress responses.

The findings suggest that maternal depression during pregnancy may result in epigenetic changes in the NR3C1 gene in neonates, which in turn may lead to alterations in cortisol stress responses in infants.

Radtke KM, Ruf M, Gunter HM, Dohrmann K, Schauer M, Meyer A, Elbert
    T. Transgenerational impact of intimate partner violence on methylation in the
    promoter of the glucocorticoid receptor. *Transl Psychiatry.* 2011 Jul 19;1(7):e21.
    doi: 10.1038/tp.2011.21. PMID: 22832523; PMCID: PMC3309516.

This article explores the epigenetic effects of intimate partner violence (IPV) on a specific gene promoter. The research indicates that IPV has a transgenerational impact, potentially affecting the DNA methylation patterns in the glucocorticoid receptor gene promoter. Methylation in this context is a chemical modification of DNA that can influence gene expression. The study may provide insights into the long-term consequences of IPV on both individuals who experience it and their descendants.

Further Reading

Daskalakis NP, Rijal CM, King C, Huckins LM, Ressler KJ. Recent genetics and
    epigenetics approaches to PTSD. *Curr Psychiatry Rep.* 2018 Apr 5;20(5):30.
    doi: 10.1007/s11920-018-0898-7. PMID: 29623448; PMCID: PMC6486832.

This article discusses recent genetics and epigenetics approaches to PTSD. It explains that genetic and epigenetic factors contribute to the development and maintenance of PTSD, with genetic risk factors including variations in several genes associated with stress response and regulation. The article also highlights epigenetic modifications such as DNA methylation and histone modifications that play a critical role in the regulation of gene expression related to PTSD. It provides evidence that epigenetic changes in specific genes associated with PTSD are influenced by environmental exposures and behavioural experiences. Finally, the article concludes that genetics and epigenetics approaches have the potential to inform new treatments and interventions for PTSD.

DeLisi M, Vaughn MG. The vindication of lamarck? Epigenetics at the intersection of
    law and mental health. *Behav Sci Law.* 2015 Oct;33(5):607–28. doi: 10.1002/
    bsl.2206. Epub 2015 Sep 21. PMID: 26387846.

The article explores the intersection of epigenetics, law and mental health, examining the implications of the emerging field of epigenetics for legal decision-making. It reviews the basic concepts of epigenetics, and how they are thought to influence behaviour and mental health, before exploring the potential implications of this research for legal concepts such as criminal responsibility and civil liability. The authors argue that epigenetics has the potential to challenge traditional notions of individual responsibility and free will, as well as to open new avenues for understanding and treating mental illness. They conclude by calling for further research into the implications of epigenetics for law and policy and for greater collaboration between the scientific and legal communities.

Zannas AS, Provençal N, Binder EB. Epigenetics of posttraumatic stress disorder: current evidence, challenges, and future directions. *Biol Psychiatry*. 2015 Sep 1;78(5):327–35. doi: 10.1016/j.biopsych.2015.04.003. Epub 2015 Apr 7. PMID: 25979620.

The review article provides an overview of current evidence, challenges and future directions in the epigenetics of PTSD. The authors discuss the role of DNA methylation, histone modifications and non-coding RNAs in PTSD pathophysiology, and highlight the importance of gene-environment interactions and intergenerational transmission of epigenetic marks. They also discuss challenges in studying epigenetic changes in PTSD, such as the need for large sample sizes, longitudinal designs and standardised methodologies. Finally, the authors suggest future directions for research, including the use of novel technologies such as single-cell sequencing and CRISPR/Cas9 gene editing to better understand the epigenetic mechanisms underlying PTSD.

8

Developing a Project

Overall, epigenetic research has the potential to significantly impact our understanding of human health and disease. Epigenetic changes have been associated with a variety of diseases, including cancer, autoimmune disorders and neurological disorders. Studying epigenetics can help us better understand the underlying pathology and develop more effective treatments and predictive tests.

8.1 Considerations for Epigenetic Research

8.1.1 What Is Meaningful in Terms of Tissue Analysed?

In terms of epigenetic research, tissue type is an important consideration. Somatic epigenetic changes are tissue-specific and dynamic, meaning multiple tissue types need to be collected at multiple time points to be able to determine cause versus consequence. In an ideal world, the relevant tissue should be collected before the onset of the disease for comparison; however, the practicalities/reality of this make this unlikely. Cancer poses slightly less of a problem because of the accessibility of the tissue after surgery/biopsy where DNA can be extracted, but cancer studies still have the issue of the heterogeneity of the tissue collected. In conditions such as Type2 Diabetes Mellitus (T2DM), the pancreatic beta cell is a sensible target cell to study to focus on the epigenetic changes associated with impaired insulin production, but currently, this can only be done in animal models. However, because T2DM is a systematic disease, it is sensible to look at multiple tissues to fully understand the epigenomic contribution to the pathogenesis associated with T2DM, comprising impaired insulin production and insulin resistance in peripheral tissues such as muscle that led to the impairment of blood glucose uptake. The feasibility/cost of studying multiple tissue types are important considerations in any project (Table 8.1).

Epigenetics and Health: A Practical Guide, First Edition. Michelle McCulley.
© 2024 John Wiley & Sons, Inc. Published 2024 by John Wiley & Sons, Inc.

Table 8.1 Tissue types and epigenetic research.

Tissue type	Advantages	Limitations
Blood	Easy to obtain, non-invasive, high cell yield	Epigenetic changes in blood may not reflect changes in other tissues, such as brain or liver
Saliva	Easy to obtain, non-invasive	Epigenetic changes in saliva may not reflect changes in other tissues and can be affected by diet and oral hygiene
Urine	Non-invasive, easy to obtain	Epigenetic changes in urine may not reflect changes in other tissues and can be affected by diet and hydration
Hair	Non-invasive, provides a long-term record of epigenetic changes	Hair may not be suitable for studying tissue-specific epigenetic changes
Skin	Easy to obtain, non-invasive, can reflect changes in environmental exposures	Epigenetic changes in the skin may not reflect changes in other tissues
Brain	Directly relevant for many neurological and psychiatric disorders	Difficult and invasive to obtain, limited availability of post-mortem tissue
Liver	Relevant for metabolic diseases	Difficult to obtain, invasive
Adipose tissue	Relevant for metabolic diseases	Difficult to obtain, invasive
Muscle	Relevant for metabolic diseases	Difficult to obtain, invasive

8.1.2   What Are the Health Promotion/Public Health Consequences?

Epigenetic changes can be influenced by environmental factors, such as pollution and toxic chemicals. Studying these changes can help us understand the impact of these factors on human health and develop strategies to minimise their negative effects. Studies to date looking at exposure to stressors and toxins appear to provide evidence for the developmental origins of health and disease (DOHaD), strengthening the link between early life exposures and later health outcomes. Important questions here are how our life experiences and the environment we exist in bring about changes in our phenotype that are then passed on to our offspring, and in some cases, particularly if the exposure is severe, may last for multiple generations. This is an important consideration for health policy decision-makers; long-term programming effects are a potential issue, so it is important to not just focus on the prenatal period but also on preconception.

There needs to be a targeted effort to investigate whether some changes to gene methylation are predictors of later disease, particularly AHRR pathway, oxidative stress, inflammation and hyothalamo-hypophyseal axis. Even if epigenetic marks are not predictors of effect, they could still be used as markers of exposure. Further research is still needed as to their reliability, specificity and effect with respect to dose, timing of exposure as well as cell and tissue specificity. There is still very much to learn in this area and currently, the research is still too preliminary to be able to recommend interventions to ameliorate disease risk based on epigenetic states. This does not mean, however, that epigenetic testing is not a potential future tool in terms of public health/health promotion initiatives. It is likely that the more we learn in this field, the better we can help to use epigenetic data to understand and reduce health risks from NCDs. What is important here is not to use this to stigmatise communities and individuals who are most affected by socioeconomic factors that increase the risk of NCDs (Table 8.2).

8.1.3   What Are the Consequences of Epigenetic Intervention?

Epigenetic research also raises ethical and policy questions around issues such as genetic privacy and the potential for epigenetic discrimination. Epigenetic modifications can be inherited, which could have implications for how traits are passed down from generation to generation. This raises important ethical questions

Table 8.2   Simplification of linkages between stress, inequity and epigenetic changes.

Factor	Description	Epigenetic changes
Stress	A physiological response to a perceived threat or challenge	Methylation of genes involved in stress response and inflammation; histone modifications leading to changes in chromatin structure; miRNA expression changes related to stress
Inequity	Unfair distribution of resources, opportunities or power	Methylation changes in genes related to stress response and inflammation; Histone modifications leading to changes in chromatin structure; DNA methylation changes in genes related to immune function and metabolism
Interactions	Stressful experiences are more likely to occur in populations that experience inequity	Gene expression changes in response to stress and inflammation; methylation changes in genes related to stress response and inflammation; DNA methylation changes in genes related to immune function and metabolism

about the long-term implications of interventions that alter epigenetic marks, such as in vitro fertilisation and gene editing, and for the clinical utility of predictive epigenetic testing. A consistent feature of mammalian ageing is DNA hypomethylation, associated with increased biological age and loss of physiologic resilience. The epigenome comprises DNA methylation and post-translational modification of chromatin-associated proteins, typically histones via acetylation, phosphorylation, ubiquitylation, sumoylation and malonylation. The epigenome also extends through noncoding RNAs that reciprocally regulate gene expression in response to environmental changes.

There are foods that could potentially influence DNA methylation. Nutrition can impact the epigenome via the epigenetic landscape of ageing and directly via providing nutritionally derived methyl donor groups to supplement the maintenance of the methylome, which is crucial to normal gene function. That is, foods rich in vitamin B12, methionine, choline, betaine and folate. Additionally, foods that affect the microbiota may also impact the epigenome and mitochondrial biogenesis, as gut microbial metabolites are taken up by the host and influence the maintenance of the epigenome. There is some evidence to suggest that certain foods and nutrients may influence DNA methylation. Some examples of foods that may potentially influence DNA methylation include folate, which is found in leafy green vegetables, beans and fortified cereals and is involved in the synthesis of *S*-adenosylmethionine (SAM), which is a key methyl donor for DNA methylation. Cruciferous vegetables such as broccoli, cauliflower and cabbage contain compounds such as sulforaphane that can influence DNA methylation. Green tea contains compounds called polyphenols, which have been shown to have epigenetic effects, including on DNA methylation. Omega-3 fatty acids, which are found in fatty fish and flaxseeds, have been shown to have epigenetic effects, as has resveratrol, which is found in red wine, grapes and berries and shown to have epigenetic effects, including on DNA methylation. Much more research is needed to fully understand the mechanisms involved and the potential health benefits or risks associated with these dietary interventions.

8.1.4 How Do We Design Epigenetic Drugs?

Epigenetic modifications can affect the expression of many different genes, which makes it difficult to target specific genes or regions of DNA without affecting others. Existing and future epigenetic drugs exert their effect through a broad range of complex and transient epigenetic modifications to the nucleosome via epigenetic proteins grouped as readers, writers and erasers. Writers catalyse the addition of epigenetic modifications to DNA and histones; readers identify and alter the interactions between protein, DNA, histone proteins and other binding factors; and erasers act as catalysts for a dynamic epigenetic response removing, the modification and permitting reversal of any changed downstream effects. The lack of specificity with epigenetic drugs can result in unwanted side effects or even harmful consequences. Epigenetic drugs can interact with proteins and other molecules in the body, leading to unintended effects on other cellular processes. These off-target effects can also cause side effects and limit the efficacy of the therapy. Like with other types of drugs, cancer cells and other diseased cells can develop resistance to epigenetic therapies over time, making it difficult to achieve long-term benefits. Because epigenetic modifications can be heritable and affect the expression of many different genes, there are ethical concerns related to the potential long-term effects of epigenetic therapy on future generations. Overall, while epigenetic therapy has shown promise in treating certain diseases, it is still a relatively new and evolving field, and further research is needed to address these limitations and fully understand its potential benefits and risks.

Because of intensive efforts focused on understanding the epigenomic contribution to cancer, the focus of novel therapeutics is also shifting towards chromatin, the rationale being that the role of chromatin is to interpret signals from growth factors and signalling molecules, and therefore targeting chromatin can mean that a more widespread action is taken, which does not always need, precise knowledge of existing mutations to be effective as a treatment. This strategy also means there is potential for activating non-genic regions of the genome, such as endogenous retroviruses, again broadening the therapeutic target area. With a broader target area, it is possible that, at least for cancer, epigenomic-targeted drugs will represent true genomic medicines in their broad effects. Traditionally, cancer drugs were developed to specifically target signals received from outside the cell, such as growth factors or hormones, or to disrupt signalling within the cytoplasm. Cytotoxic drugs that induce DNA damage or interfere with mitosis in the nucleus form the basis of most chemotherapy treatments. The development of drugs to target chromatin is a different approach but is exciting as it has the potential to produce less toxic treatments that target specific enzymes. Also, targeting chromatin can enhance the activities of other drugs delivered in combination therapies.

Currently, there are, broadly speaking, two main categories of epigenomically targeted drugs that are currently in clinical trial: broad reprogrammers and drugs developed to treat more specific subsets of patients, which represents more typical precision medicine in terms of a targeted therapy. Examples of broad reprogrammers include DNMT inhibitors (DNMTi), histone deacetylase inhibitors (HDACi) and inhibitors of the bromodomain and extra-terminal motif proteins (iBETS). All these types of treatments intend to bring about large-scale changes in gene expression, typically aiming to reverse cancer-specific gene expression changes. The use of DNMTi has produced some satisfactory results with long-term responses in patients; limitations to date appear to be the development of resistance to the therapies, although second-generation hypomethylating drugs seem to be having better results. Some HDACi drugs have been approved for some malignancies and appear most effective when used in combination with other treatments. iBETS targets BRD4, a gene translocated in some cancers that reads the acetylated histone signal necessary for high-level expression of oncogenes such as MYC through the promotion of enhancer activity.

In terms of targeted therapies, the H3K27 histone *N*-methyltransferase EZH2 is found to be activated by mutations in lymphomas, so using an EZH2 inhibitor is designed to induce selective killing of cells carrying this mutation. First-generation IDH inhibitors designed to inhibit DNA and histone demethylation have shown some success in trial but remains to be seen whether they can be as effective as DNMT inhibition. Another strategy is targeting synthetic lethality – this refers to the relationship between two genes whereby their combined inactivation results in cell death but not their individual activation, or a gene whereby cell death only eventuates if there is a particular cellular feature. Drugs that inhibit the H3K79 N-methyltransferase DOT1L appear active in vitro in leukaemias with activation of KMT2A.

Epigenomic therapies as outlined above rely on restoring the activity of genes that have become abnormally silenced during carcinogenesis. There is also the possibility of activating genes and repetitive DNA elements that are repressed in normal and cancer cells in an attempt to enhance the patient's response to therapy. Activation of the host immune defence mechanisms through increased expression of tumour antigens would help increase the visibility of tumours to the host immune defence mechanisms. Similarly, the activation of endogenous retroviruses can also increase immunogenicity. Such treatments can be targeted by focusing the drug targets on S-phase cells, as most cells in the human body are quiescent.

DNMTi therefore has a widespread effect on both abnormally silenced genes relevant to cancer and on genes and repetitive DNAs held silenced by epigenetic marks. If a drug inhibits global DNA methylation then the overall effect would be to reset the epigenome and activate several pathways simultaneously, potentially increasing the efficacy of these drugs, this has therapeutic potential. Most pathways silenced epigenetically during tumour development would have been selected based on conferring a growth/survival advantage to the tumour, making a negative effect unlikely, though possible.

CRISPR/Cas technology can be used in epigenetic therapy to modify or edit the epigenetic marks on DNA, which can potentially treat or prevent disease. This is known as epigenome editing, and it involves using the Cas enzyme and guide RNA to target and modify specific epigenetic marks, such as DNA methylation or histone modifications. By editing these marks, it is possible to turn genes on or off, alter their expression levels or modify the way they interact with other genes and cellular processes. This can be useful for treating or preventing a variety of diseases, including genetic disorders, cancer and neuro-degenerative diseases. One potential application of epigenome editing is to reactivate genes that have been silenced by epigenetic marks, which can be useful for treating diseases such as cancer or genetic disorders. Another application is to introduce new epigenetic marks to modify the expression of genes, which can be useful for developing new treatments for a variety of diseases. While epigenome editing using CRISPR/Cas is a promising approach, more research is needed to fully understand the potential benefits and risks of this technology in the context of epigenetic therapy. Some of the ethical concerns related to CRISPR/Cas, such as off-target effects and heritability, also apply to its use in epigenetic therapy.

8.1.5 What Does Predictive Epigenetic Testing Look Like?

Predictive epigenetic testing is a type of genetic testing that analyses a person's epigenetic marks to predict their risk of developing certain diseases or conditions. Based on the results, the test can provide information about a person's risk of developing the disease or condition in question. Predictive epigenetic testing has the potential to provide valuable information for disease prevention, early detection and personalised treatment. It also raises ethical concerns related to privacy, discrimination and psychological impact. A key strength the earlier identification of people who are at greater risk of non-communicable diseases, which as such can be used as a health promotion tool to aid in motivation/strategies to improve/increase preventative measures. For example, epigenetic biomarkers are indicative of increased cardiovascular disease. This could also be useful in cancer precision medicine strategies to identify drugs that are likely to be more effective in an individual based on their cancer profile. Obviously, this would be time and resource heavy and there is still the issue with health promotion strategies in that to encourage behaviour change, individuals must perceive they are at risk and that the data is informative, valuable, trustworthy, etc. So, an important consideration here is how to ensure epigenetic testing does provide clinical use and is likely to impact positively on public health.

Pros	Cons
May identify potential health risks and diseases	Results may cause anxiety or stress
Can lead to early detection and intervention	Results may lead to stigmatisation or discrimination

(Continued)

Pros	Cons
May help inform lifestyle changes or interventions	Predictive testing is not always accurate or reliable
Can inform family planning decisions	Limited knowledge of how epigenetic changes relate to health
Could contribute to personalised medicine	Cost of testing may be prohibitive
May inform disease prevention strategies	Lack of regulatory oversight for epigenetic testing
Can provide valuable information for research	Results may have limited clinical utility

8.2 Conclusion

It is clear there is still much to understand in the field of epigenetics. What is becoming increasingly clearer is the need to approach understanding epigenetic regulation from multiple levels simultaneously. Future understanding of epigenetics is going to need the integration of the different layers of epigenetic information from DNA methylation and posttranslational modifications of histones to the restructuring of chromatin and transcriptomic changes. There are, however, limitations of epigenetic therapy; these include lack of specificity, off-target effects, resistance and ethical issues.

The philosophy of epigenetics centres around the idea that genes are not the sole determinants of an organism's traits and development and that the environment, including social and psychological factors, can also play a role in shaping an individual's phenotype. Epigenetics challenges the traditional view of genetics as a deterministic, linear process in which genes dictate the development and behaviour of an organism. Instead, it suggests that gene expression is a dynamic, complex process that is influenced by a variety of factors, including environmental cues, interactions with other genes and epigenetic modifications. Epigenetics also raises important philosophical questions about the nature of inheritance, identity and agency. For example, if epigenetic marks can be passed on to future generations, how does this affect our understanding of inheritance and the role of DNA in shaping identity? If epigenetic modifications can influence behaviour, can individuals be held fully responsible for their actions? It is important we keep raising these questions. Epigenetics emphasises the importance of understanding the complex interplay between genes and the environment and challenges us to think more critically about the nature of biological processes, identity and agency, arguably science as it should be.

8.3 Getting Started

Why Do Epigenetic Research? What Do You Want to Achieve?

The potential application of epigenetic biomarkers is huge in that they may be utilised to identify and treat aberrant methylation patterns long before the onset of symptoms. Whilst there is still a long way to go for us to understand epigenetics, it is useful to frame your research question in a broader context to focus on the endpoint of why you want to research a particular area, albeit better understanding, better treatment, better environmental/population outcomes, etc. (Table 8.3).

Step 1
1) What is your research question? What are you hoping to find out? Why are you doing this? How is it going to benefit humanity?
2) What has already been done in this area, and what is already known? Are there people currently in this field who would be good collaborators? Are there existing publicly available data sets that you can use to draw from for preliminary analysis? Look at current resources, e.g. TaRGET II programme, US National Institutes of Health Roadmap, Epigenomics Consortium
3) What is your research question going to add to the current knowledge on this topic/what gap is it going to fill?

Step 2
1) Who/what is your target population, and how are you going to recruit/collect samples?
2) What are the ethical issues/permissions needed to research your target population is this achievable/realistic?
3) Do samples need to be collected at multiple timepoints? What are the logistical, administrative and cost implications for this? How are these time intervals decided? What evidence is there to support that these time points will be meaningful? Distinguishing between age-related change and exposure-mediated change is difficult in humans, with limited levels of variable control, inherent cohort bias, and missing exposure data. Difficult to assess the effects of early life exposure on long-term epigenetic changes, not possible to determine whether age-related diseases are the cause or effect of epigenetic changes.

Step 3
1) What tissue is most meaningful to your research question?
   The biggest challenge is in determining and collecting the most appropriate tissue to be studied for each complex disease trait. Tissue heterogeneity and availability. Feasibility in terms of sample collection, cost-effectiveness and time to study all tissue types.

Table 8.3 Summary flow diagram of steps involved in planning an epigenetics research project.

Step 1

- Determine the research question, aims and potential benefits to humanity
- Review existing literature and resources
- Identify the research gap that the project will address

Step 2

- Identify target population and recruitment methods
- Consider ethical issues and permissions needed
- Determine whether samples need to be collected at multiple timepoints

Step 3

- Determine the most appropriate tissue(s) to study
- Consider sample size and study design
- Consider the use of stem cells

Step 4

- Address ethical issues associated with the research
- Identify appropriate authorities to receive ethical approval

Step 5

- Determine appropriate epigenetic modifications to investigate
- Consider available methods and funding
- Determine the realistic scale of the project

Step 6

- Identify laboratory equipment needed
- Decide whether to outsource certain steps of the project

Step 7

- Determine appropriate software and computer skills needed for data analysis
- Consider upskilling or outsourcing this aspect

Step 8

- Identify potential funding sources for the project

Step 9

- Determine a justifiable sample size for the study
- Consider the statistical power needed to detect significant epigenetic changes

Step 10

- Write a sample study proposal with costs, timelines and resources needed
- Adjust the project scale based on available funding

Step 11

- Begin writing the research project proposal
- Apply for ethical approval and funding

2) Can disease-associated epigenetic changes be revealed by simple case-control study design, or whether more robust study designs need to be developed?
3) What sample size is required to achieve statistical power to detect a disease-associated change?
4) Consider the use of stem cells for ageing epigenome studies.

Step 4
1) What are the ethical issues associated with your research?
2) Who do you need to approach to receive ethical approval for your project?

Step 5
1) What methods are most applicable to your research question? Are you looking at DNA methylation or histone modifications? Also need to explore other epigenetic modifications alongside DNA methylation, e.g. using ATAC sequencing to investigate chromatin state. Look at 5-hmC – traditional bisulphite sequencing does not distinguish between 5-hmC and 5-mC, but newer technologies such as hMeDIP-seq can measure genome-wide 5-hmC levels.5hmC stable epigenetic mark enriched at transcription factor binding sites, enhancers and genes but depleted at promoters, suggesting a complex role in gene regulation. Global loss of 5-hmC associated with cancer development.
2) What is a realistic scale for your project given the methods you want to use and the funds you are likely to have available?

Step 6
1) How much of the analysis are you equipped to do in your laboratory? What equipment do you need?
2) Are you going to outsource steps, e.g. library prep, sequencing and analysis of data?

Step 7
1) How are you going to analyse the results?
2) Do you have appropriate computer/software/skills – how are you going to upskill or outsource this aspect?

Step 8
1) What funding is available to apply for in order to carry out your project?

Step 9
1) What is a justifiable sample size for your study?
   Are simple case-control studies sufficient to detect significant epigenetic changes, or what sample size would be required to achieve sufficient statistical power? It may be the case that more robust study designs need to be developed. Tsai and Bell (2015) published data on power and sample size estimation for EWAS to detect differential methylation. They report that a study comprising

25 case-controls has 45% power to detect an 8% methylation difference genome-wide at $p = 0.05$ significance, but 85% to detect significance at a single locus. As with all power calculations, the effect size is also important in determining the sample size, with a greater mean methylation difference increasing the power. With a study of 17 case controls, we would expect 80% power to detect an 8% methylation difference at a single locus with significance $p = 0.05$.

Step 10
1) Write a sample study proposal, including costs, timelines and resources.
2) Is it realistic? Adjust accordingly/scale up/down your project to meet the funding amount.

Step 11
1) Start writing,
2) Apply for funding and ethical approval.

Reference

Tsai PC, Bell JT. Power and sample size estimation for epigenome-wide association scans to detect differential DNA methylation. *Int J Epidemiol.* 2015 Aug;44(4): 1429–41. doi: 10.1093/ije/dyv041. Epub 2015 May 13. PMID: 25972603; PMCID: PMC4588864.

This article discusses the importance of power and sample size estimation in epigenome-wide association scans (EWAS) to detect differential DNA methylation. The authors provide an overview of statistical methods for estimating power and sample size in EWAS and highlight the factors that influence the power of these studies, including effect size, sample size and the number of CpG sites analysed. The authors also discuss the use of simulation studies and power analysis tools to improve the design and interpretation of EWAS. They conclude that careful consideration of power and sample size is critical for the success of EWAS and provide guidance on how to optimise these factors in future studies.

Further Reading

Goell JH, Hilton IB. CRISPR/Cas-based epigenome editing: advances, applications, and clinical utility. *Trends Biotechnol.* 2021 Jul;39(7):678–91. doi: 10.1016/j. tibtech.2020.10.012. Epub 2021 May 7. PMID: 33972106.

The review article 'CRISPR/Cas-Based Epigenome Editing: Advances, Applications and Clinical Utility' by Goell and Hilton provides an overview of recent advances in epigenome editing using CRISPR/Cas technology. The authors discuss the potential applications of epigenome editing, such as in the treatment of genetic diseases, cancer therapy and precision medicine. The review covers the different epigenetic modifications that can be targeted using CRISPR/Cas technology, including DNA methylation and histone modifications. The authors also discuss the challenges and limitations of epigenome editing, such as off-target effects and the need for efficient and specific delivery methods. The article concludes by highlighting the promising future of epigenome editing as a powerful tool for understanding gene regulation and developing new therapeutic strategies for various diseases. Overall, the article provides a comprehensive overview of the current state and potential of CRISPR/Cas-based epigenome editing.

Holtzman L, Gersbach CA. Editing the epigenome: reshaping the genomic landscape. *Annu Rev Genomics Hum Genet.* 2018 Aug 31; 19:43–71. doi: 10.1146/annurev-genom-083117-021632. Epub 2018 May 31. PMID: 29852072.

The review article provides an overview of epigenetic modifications and their role in regulating gene expression. The authors discuss recent advances in epigenome editing technologies such as CRISPR/Cas9, TALEs and zinc finger nucleases, and their potential applications in basic research, drug discovery and gene therapy. The review also covers the ethical and societal implications of epigenome editing and the challenges that need to be addressed for safe and effective clinical use. Overall, the article highlights the potential of epigenome editing as a powerful tool for investigating gene function and developing new therapeutic strategies for genetic diseases.

Martin CL, Ghastine L, Lodge EK, Dhingra R, Ward-Caviness CK. Understanding health inequalities through the lens of social epigenetics. *Annu Rev Public Health* 2022;43:235–54. doi: 10.1146/annurev-publhealth-052020-105613. PMID: 35380065; PMCID: PMC9584166.

The article discusses how social epigenetics can provide a framework for understanding health inequalities. The authors highlight the impact of social determinants of health, including socioeconomic status, race and environmental factors, on epigenetic modifications and gene expression. They provide examples of how these factors can lead to differential health outcomes and highlight the potential for interventions that target social determinants of health to improve health equity. The article also discusses the challenges and limitations of studying social epigenetics, including issues related to data quality and interpretation. The authors conclude by highlighting the need for interdisciplinary collaboration to further advance the field of social epigenetics and reduce health inequalities. Overall, the article provides valuable insights into the complex relationship between social determinants of health, epigenetics and health outcomes.

Nagy C, Turecki G. Transgenerational epigenetic inheritance: an open discussion. *Epigenomics*. 2015 Aug;7(5):781–90. doi: 10.2217/epi.15.46. Epub 2015 Sep 7. PMID: 26344807.

The review article 'Transgenerational epigenetic inheritance: an open discussion' by Nagy and Turecki examines the current understanding and controversies surrounding the transmission of epigenetic modifications from one generation to the next. The authors discuss the evidence for epigenetic inheritance in humans and animal models, including studies on the effects of parental experiences, such as stress and trauma, on the epigenetic profiles of offspring. The review also covers the mechanisms underlying transgenerational epigenetic inheritance, such as DNA methylation and histone modifications, and the challenges in studying this phenomenon, including the difficulty in distinguishing between genetic and epigenetic inheritance. The authors conclude by calling for more interdisciplinary research and open discussion to further understand and address the implications of transgenerational epigenetic inheritance for human health and disease.

Santaló J, Berdasco M. Ethical implications of epigenetics in the era of personalized medicine. *Clin Epigenetics*. 2022 Mar 25;14(1):44. doi: 10.1186/s13148-022-01263-1. PMID: 35337378; PMCID: PMC8953972.

The article explores the ethical implications of epigenetics in the context of personalised medicine, which uses genomic and epigenomic information to tailor medical interventions to individual patients. The authors discuss issues related to privacy, informed consent, genetic discrimination and equity in access to personalised medicine. They also highlight the potential benefits of personalised medicine, such as improved disease prevention and treatment outcomes, and the need to balance these benefits with ethical considerations. The authors emphasise the importance of developing guidelines and policies to address the ethical challenges of epigenetics and personalised medicine, and the need for ongoing dialogue between stakeholders, including patients, healthcare providers, researchers and policymakers.

Index

Please note: Page numbers in **Bold** refer to entries in tables.

Epigenetics and Health: A Practical Guide, First Edition. Michelle McCulley.
© 2024 John Wiley & Sons, Inc. Published 2024 by John Wiley & Sons, Inc.

Printed and bound by CPI Group (UK) Ltd, Croydon, CR0 4YY

27/10/2024

14580476-0003